Student Edition

Eureka Math
Grade 4
Modules 3 & 4

Special thanks go to the Gordon A. Cain Center and to the Department of Mathematics at Louisiana State University for their support in the development of *Eureka Math*.

For a free *Eureka Math* Teacher Resource Pack, Parent Tip Sheets, and more please visit www.Eureka.tools

Published by the non-profit Great Minds

Copyright © 2015 Great Minds. No part of this work may be reproduced, sold, or commercialized, in whole or in part, without written permission from Great Minds. Non-commercial use is licensed pursuant to a Creative Commons Attribution-NonCommercial-ShareAlike 4.0 license; for more information, go to http://greatminds.net/maps/math/copyright. "Great Minds" and "Eureka Math" are registered trademarks of Great Minds.

Printed in the U.S.A.

This book may be purchased from the publisher at eureka-math.org

10 9

ISBN 978-1-63255-304-1

Name _____   Date _____

1. Determine the perimeter and area of rectangles A and B.

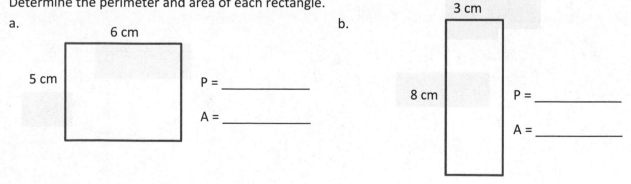

    a. A = _____   A = _____

    b. P = _____   P = _____

2. Determine the perimeter and area of each rectangle.

    a.

6 cm

5 cm

P = _____

A = _____

    b.

3 cm

8 cm

P = _____

A = _____

EUREKA MATH™

©2015 Great Minds. eureka-math.org
G4-M3-M4-SE-1.3.1-01.2016

3. Determine the perimeter of each rectangle.

a.

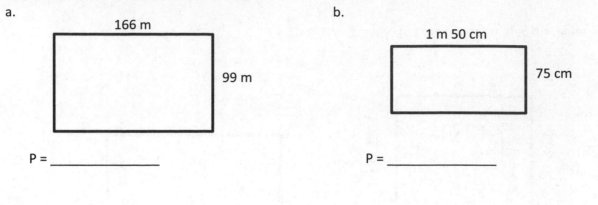

166 m

99 m

P = _____

b.

1 m 50 cm

75 cm

P = _____

4. Given the rectangle's area, find the unknown side length.

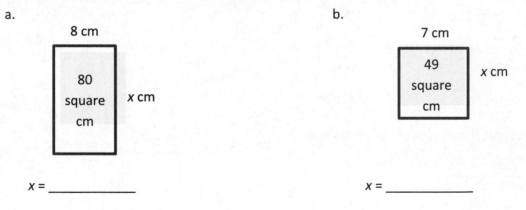

a.

8 cm

80 square cm

x cm

x = _____

b.

7 cm

49 square cm

x cm

x = _____

Lesson 1:   Investigate and use the formulas for area and perimeter of rectangles.

EUREKA MATH™

©2015 Great Minds. eureka-math.org
G4-M3-M4-SE-1.3.1-01.2016

5. Given the rectangle's perimeter, find the unknown side length.

a.   P = 120 cm

b.   P = 1,000 m

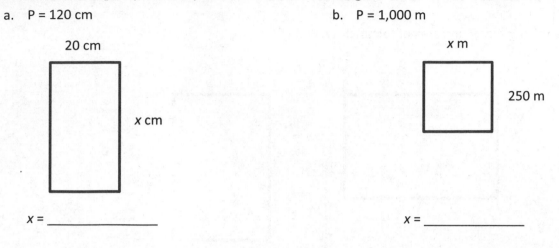

20 cm

x cm

x = _____

x m

250 m

x = _____

6. Each of the following rectangles has whole number side lengths. Given the area and perimeter, find the length and width.

a.   P = 20 cm

b.   P = 28 m

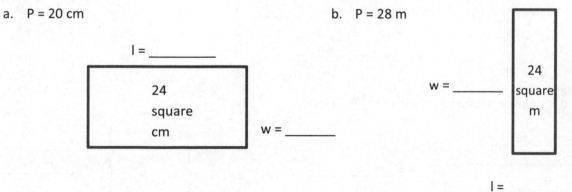

l = _____

24 square cm

w = _____

24 square m

w = _____

l = _____

EUREKA MATH™

Lesson 1:   Investigate and use the formulas for area and perimeter of rectangles.

3

©2015 Great Minds. eureka-math.org
G4-M3-M4-SE-1.3.1-01.2016

Name _____   Date _____

1. Determine the perimeter and area of rectangles A and B.

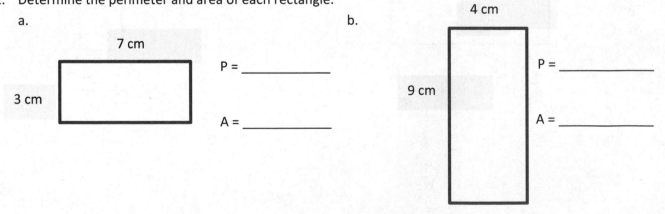

a. A = _____   A = _____

b. P = _____   P = _____

2. Determine the perimeter and area of each rectangle.

a.

7 cm

3 cm

P = _____

A = _____

b.

4 cm

9 cm

P = _____

A = _____

Lesson 1:   Investigate and use the formulas for area and perimeter of rectangles.

©2015 Great Minds. eureka-math.org
G4-M3-M4-SE-1.3.1-01.2016

3. Determine the perimeter of each rectangle.

a.

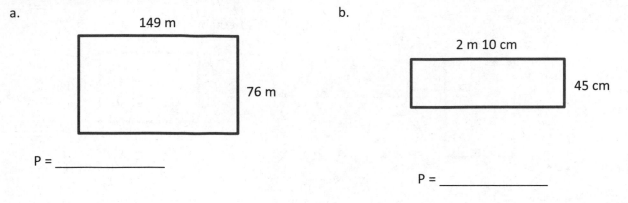

149 m

76 m

P = _____

b.

2 m 10 cm

45 cm

P = _____

4. Given the rectangle's area, find the unknown side length.

a.

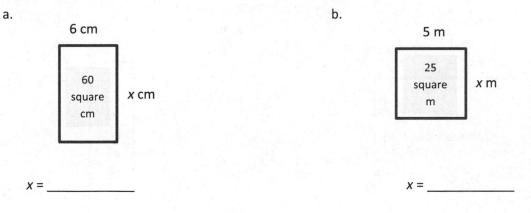

6 cm

60 square cm

x cm

x = _____

b.

5 m

25 square m

x m

x = _____

5. Given the rectangle's perimeter, find the unknown side length.

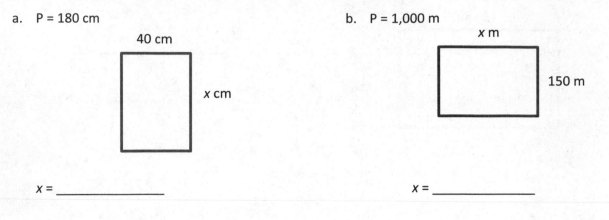

a. P = 180 cm

40 cm

x cm

x = _____

b. P = 1,000 m

x m

150 m

x = _____

6. Each of the following rectangles has whole number side lengths. Given the area and perimeter, find the length and width.

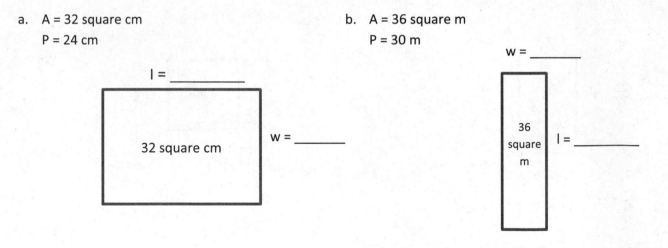

a. A = 32 square cm
P = 24 cm

l = _____

32 square cm

w = _____

b. A = 36 square m
P = 30 m

w = _____

36 square m

l = _____

Lesson 1:   Investigate and use the formulas for area and perimeter of rectangles.

©2015 Great Minds. eureka-math.org
G4-M3-M4-SE-1.3.1-01.2016

Name _____   Date _____

1. A rectangular porch is 4 feet wide. It is 3 times as long as it is wide.

   a. Label the diagram with the dimensions of the porch.

   b. Find the perimeter of the porch.

2. A narrow rectangular banner is 5 inches wide. It is 6 times as long as it is wide.

   a. Draw a diagram of the banner, and label its dimensions.

   b. Find the perimeter and area of the banner.

3. The area of a rectangle is 42 square centimeters. Its length is 7 centimeters.

   a. What is the width of the rectangle?

   b. Charlie wants to draw a second rectangle that is the same length but is 3 times as wide. Draw and label Charlie's second rectangle.

   c. What is the perimeter of Charlie's second rectangle?

Lesson 2:   Solve multiplicative comparison word problems by applying the area and perimeter formulas.

©2015 Great Minds. eureka-math.org
G4-M3-M4-SE-1.3.1-01.2016

4. The area of Betsy's rectangular sandbox is 20 square feet. The longer side measures 5 feet. The sandbox at the park is twice as long and twice as wide as Betsy's.

    a. Draw and label a diagram of Betsy's sandbox. What is its perimeter?

    b. Draw and label a diagram of the sandbox at the park. What is its perimeter?

    c. What is the relationship between the two perimeters?

    d. Find the area of the park's sandbox using the formula $A = l \times w$.

EUREKA MATH

Lesson 2: Solve multiplicative comparison word problems by applying the area and perimeter formulas.

©2015 Great Minds. eureka-math.org
G4-M3-M4-SE-1.3.1-01.2016

9

e. The sandbox at the park has an area that is how many times that of Betsy's sandbox?

f. Compare how the perimeter changed with how the area changed between the two sandboxes. Explain what you notice using words, pictures, or numbers.

Lesson 2: Solve multiplicative comparison word problems by applying the area and perimeter formulas.

©2015 Great Minds. eureka-math.org
G4-M3-M4-SE-1.3.1-01.2016

Name _____ Date _____

1. A rectangular pool is 7 feet wide. It is 3 times as long as it is wide.

    a. Label the diagram with the dimensions of the pool.

    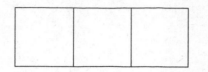

    b. Find the perimeter of the pool.

2. A poster is 3 inches long. It is 4 times as wide as it is long.

    a. Draw a diagram of the poster, and label its dimensions.

    b. Find the perimeter and area of the poster.

Lesson 2:   Solve multiplicative comparison word problems by applying the area
             and perimeter formulas.

©2015 Great Minds. eureka-math.org
G4-M3-M4-SE-1.3.1-01.2016

11

3. The area of a rectangle is 36 square centimeters, and its length is 9 centimeters.

    a. What is the width of the rectangle?

    b. Elsa wants to draw a second rectangle that is the same length but is 3 times as wide. Draw and label Elsa's second rectangle.

    c. What is the perimeter of Elsa's second rectangle?

Lesson 2:   Solve multiplicative comparison word problems by applying the area and perimeter formulas.

©2015 Great Minds. eureka-math.org
G4-M3-M4-SE-1.3.1-01.2016

4. The area of Nathan's bedroom rug is 15 square feet. The longer side measures 5 feet. His living room rug is twice as long and twice as wide as the bedroom rug.

a. Draw and label a diagram of Nathan's bedroom rug. What is its perimeter?

b. Draw and label a diagram of Nathan's living room rug. What is its perimeter?

c. What is the relationship between the two perimeters?

d. Find the area of the living room rug using the formula A = l × w.

Lesson 2:   Solve multiplicative comparison word problems by applying the area and perimeter formulas.

13

©2015 Great Minds. eureka-math.org
G4-M3-M4-SE-1.3.1-01.2016

e. The living room rug has an area that is how many times that of the bedroom rug?

f. Compare how the perimeter changed with how the area changed between the two rugs. Explain what you notice using words, pictures, or numbers.

Lesson 2: Solve multiplicative comparison word problems by applying the area and perimeter formulas.

©2015 Great Minds. eureka-math.org
G4-M3-M4-SE-1.3.1-01.2016

Name _____ Date _____

Solve the following problems. Use pictures, numbers, or words to show your work.

1. The rectangular projection screen in the school auditorium is 5 times as long and 5 times as wide as the rectangular screen in the library. The screen in the library is 4 feet long with a perimeter of 14 feet. What is the perimeter of the screen in the auditorium?

2. The width of David's rectangular tent is 5 feet. The length is twice the width. David's rectangular air mattress measures 3 feet by 6 feet. If David puts the air mattress in the tent, how many square feet of floor space will be available for the rest of his things?

EUREKA
MATH™

Lesson 3: Demonstrate understanding of area and perimeter formulas by solving multi-step real-world problems.

15

©2015 Great Minds. eureka-math.org
G4-M3-M4-SE-1.3.1-01.2016

3. Jackson's rectangular bedroom has an area of 90 square feet. The area of his bedroom is 9 times that of his rectangular closet. If the closet is 2 feet wide, what is its length?

4. The length of a rectangular deck is 4 times its width. If the deck's perimeter is 30 feet, what is the deck's area?

Lesson 3: Demonstrate understanding of area and perimeter formulas by solving multi-step real-world problems.

©2015 Great Minds. eureka-math.org
G4-M3-M4-SE-1.3.1-01.2016

Name _____ Date _____

Solve the following problems. Use pictures, numbers, or words to show your work.

1. Katie cut out a rectangular piece of wrapping paper that was 2 times as long and 3 times as wide as the box that she was wrapping. The box was 5 inches long and 4 inches wide. What is the perimeter of the wrapping paper that Katie cut?

2. Alexis has a rectangular piece of red paper that is 4 centimeters wide. Its length is twice its width. She glues a rectangular piece of blue paper on top of the red piece measuring 3 centimeters by 7 centimeters. How many square centimeters of red paper will be visible on top?

Lesson 3: Demonstrate understanding of area and perimeter formulas by solving multi-step real-world problems.

17

©2015 Great Minds. eureka-math.org
G4-M3-M4-SE-1.3.1-01.2016

3. Brinn's rectangular kitchen has an area of 81 square feet. The kitchen is 9 times as many square feet as Brinn's pantry. If the rectangular pantry is 3 feet wide, what is the length of the pantry?

4. The length of Marshall's rectangular poster is 2 times its width. If the perimeter is 24 inches, what is the area of the poster?

Lesson 3:   Demonstrate understanding of area and perimeter formulas by solving multi-step real-world problems.

©2015 Great Minds. eureka-math.org
G4-M3-M4-SE-1.3.1-01.2016

Name _____   Date _____

Example:

$5 \times 10 =$ __50__

5 ones $\times 10 =$ __5__ __tens__

thousands	hundreds	tens	ones

Draw place value disks and arrows as shown to represent each product.

1. $5 \times 100 =$ _____

     $5 \times 10 \times 10 =$ _____

     5 ones $\times 100 =$ ____ _____

thousands	hundreds	tens	ones

2. $5 \times 1,000 =$ _____

     $5 \times 10 \times 10 \times 10 =$ _____

     5 ones $\times 1,000 =$ ____ _____

thousands	hundreds	tens	ones

3.   Fill in the blanks in the following equations.

   a. $6 \times 10 =$ _____   b.   _____ $\times 6 = 600$ c. $6,000 =$ _____ $\times 1,000$

   d. $10 \times 4 =$ _____   e. $4 \times$ _____ $= 400$ f.   _____ $\times 4 = 4,000$

   g. $1,000 \times 9 =$ _____   h.   _____ $= 10 \times 9$ i. $900 =$ _____ $\times 100$

Draw place value disks and arrows to represent each product.

4. 12 × 10 = _____

    (1 ten 2 ones) × 10 = _____

thousands	hundreds	tens	ones

5. 18 × 100 = _____

    18 × 10 × 10 = _____

    (1 ten 8 ones) × 100 = _____

thousands	hundreds	tens	ones

6. 25 × 1,000 = _____

    25 × 10 × 10 × 10 = _____

    (2 tens 5 ones) × 1,000 =

    _____

ten thousands	thousands	hundreds	tens	ones

Decompose each multiple of 10, 100, or 1,000 before multiplying.

7. 3 × 40 = 3 × 4 × _____

    = 12 × _____

    = _____

8. 3 × 200 = 3 × _____ × _____

    = _____ × _____

    = _____

9. 4 × 4,000 = _____ × _____ × _____

    = _____ × _____

    = _____

10. 5 × 4,000 = _____ × _____ × _____

    = _____ × _____

    = _____

©2015 Great Minds. eureka-math.org
G4-M3-M4-SE-1.3.1-01.2016

EUREKA
MATH™

Name _____ Date _____

Example:

$5 \times 10 =$ ___50___

5 ones $\times 10 =$ _5_ _tens_____

thousands	hundreds	tens	ones

Draw place value disks and arrows as shown to represent each product.

1. $7 \times 100 =$ _____

    $7 \times 10 \times 10 =$ _____

    7 ones $\times 100 =$ _____

thousands	hundreds	tens	ones

2. $7 \times 1,000 =$ _____

    $7 \times 10 \times 10 \times 10 =$ _____

    7 ones $\times 1,000 =$ ____

    _____

thousands	hundreds	tens	ones

3. Fill in the blanks in the following equations.

    a. $8 \times 10 =$ _____

    b. _____ $\times 8 = 800$

    c. $8,000 =$ _____ $\times 1,000$

    d. $10 \times 3 =$ _____

    e. $3 \times$ _____ $= 3,000$

    f. _____ $\times 3 = 300$

    g. $1,000 \times 4 =$ _____

    h. _____ $= 10 \times 4$

    i. $400 =$ _____ $\times 100$

Lesson 4: Interpret and represent patterns when multiplying by 10, 100, and 1,000 in arrays and numerically.

©2015 Great Minds. eureka-math.org
G4-M3-M4-SE-1.3.1-01.2016

21

Draw place value disks and arrows to represent each product.

4. 15 × 10 = _____

    (1 ten 5 ones) × 10 = _____

thousands	hundreds	tens	ones

5. 17 × 100 = _____

    17 × 10 × 10 = _____

    (1 ten 7 ones) × 100 = _____

thousands	hundreds	tens	ones

6. 36 × 1,000 = _____

    36 × 10 × 10 × 10 = _____

    (3 tens 6 ones) × 1,000 = _____

ten thousands	thousands	hundreds	tens	ones

Decompose each multiple of 10, 100, or 1000 before multiplying.

7. 2 × 80 = 2 × 8 × _____

    = 16 × _____

    = _____

8. 2 × 400 = 2 × _____ × _____

    = _____ × _____

    = _____

9. 5 × 5,000 = _____ × _____ × _____

    = _____ × _____

    = _____

10. 7 × 6,000 = _____ × _____ × _____

    = _____ × _____

    = _____

Lesson 4: Interpret and represent patterns when multiplying by 10, 100, and 1,000 in arrays and numerically.

©2015 Great Minds. eureka-math.org
G4-M3-M4-SE-1.3.1-01.2016

EUREKA MATH™

thousands	hundreds	tens	ones

thousands place value chart

Lesson 4:   Interpret and represent patterns when multiplying by 10, 100, and
1,000 in arrays and numerically.

23

©2015 Great Minds. eureka-math.org
G4-M3-M4-SE-1.3.1-01.2016

This page intentionally left blank

Name _____ Date _____

Draw place value disks to represent the value of the following expressions.

1. 2 × 3 = _____

    2 times _____ ones is _____ ones.

thousands	hundreds	tens	ones

    $$\begin{array}{r} 3 \\ \times\ \ 2 \\ \hline \end{array}$$

2. 2 × 30 = _____

    2 times _____ tens is _____ .

thousands	hundreds	tens	ones

    $$\begin{array}{r} 30 \\ \times\ \ 2 \\ \hline \end{array}$$

3. 2 × 300 = _____

    2 times _____ is _____.

thousands	hundreds	tens	ones

    $$\begin{array}{r} 300 \\ \times\ \ 2 \\ \hline \end{array}$$

4. 2 × 3,000 = _____

    ____ times _____ is _____ .

thousands	hundreds	tens	ones

    $$\begin{array}{r} 3,000 \\ \times\ \ 2 \\ \hline \end{array}$$

EUREKA MATH

Lesson 5: Multiply multiples of 10, 100, and 1,000 by single digits, recognizing patterns.

©2015 Great Minds. eureka-math.org
G4-M3-M4-SE-1.3.1-01.2016

25

5. Find the product.

a. 20 × 7	b. 3 × 60	c. 3 × 400	d. 2 × 800
e. 7 × 30	f. 60 × 6	g. 400 × 4	h. 4 × 8,000
i. 5 × 30	j. 5 × 60	k. 5 × 400	l. 8,000 × 5

6. Brianna buys 3 packs of balloons for a party. Each pack has 60 balloons. How many balloons does Brianna have?

Lesson 5: Multiply multiples of 10, 100, and 1,000 by single digits, recognizing patterns.

©2015 Great Minds. eureka-math.org
G4-M3-M4-SE-1.3.1-01.2016

7. Jordan has twenty times as many baseball cards as his brother. His brother has 9 cards. How many cards does Jordan have?

8. The aquarium has 30 times as many fish in one tank as Jacob has. The aquarium has 90 fish. How many fish does Jacob have?

Lesson 5:   Multiply multiples of 10, 100, and 1,000 by single digits, recognizing
              patterns.

©2015 Great Minds. eureka-math.org
G4-M3-M4-SE-1.3.1-01.2016

27

Name _____ Date _____

Draw place value disks to represent the value of the following expressions.

1. 5 × 2 = _____

   5 times _____ ones is _____ ones.

thousands	hundreds	tens	ones

   ```
       2
   ×   5
   ─────
   ```

2. 5 × 20 = _____

   5 times _____ tens is _____.

thousands	hundreds	tens	ones

   ```
      2 0
   ×    5
   ──────
   ```

3. 5 × 200 = _____

   5 times _____ is _____.

thousands	hundreds	tens	ones

   ```
      2 0 0
   ×      5
   ────────
   ```

4. 5 × 2,000 = _____

   _____ times _____ is _____.

thousands	hundreds	tens	ones

   ```
      2,0 0 0
   ×        5
   ──────────
   ```

Lesson 5:   Multiply multiples of 10, 100, and 1,000 by single digits, recognizing patterns.

©2015 Great Minds. eureka-math.org
G4-M3-M4-SE-1.3.1-01.2016

5. Find the product.

a. 20 × 9	b. 6 × 70	c. 7 × 700	d. 3 × 900
e. 9 × 90	f. 40 × 7	g. 600 × 6	h. 8 × 6,000
i. 5 × 70	j. 5 × 80	k. 5 × 200	l. 6,000 × 5

6. At the school cafeteria, each student who orders lunch gets 6 chicken nuggets. The cafeteria staff prepares enough for 300 kids. How many chicken nuggets does the cafeteria staff prepare altogether?

Lesson 5: Multiply multiples of 10, 100, and 1,000 by single digits, recognizing patterns.

©2015 Great Minds. eureka-math.org
G4-M3-M4-SE-1.3.1-01.2016

29

7. Jaelynn has 30 times as many stickers as her brother. Her brother has 8 stickers. How many stickers does Jaelynn have?

8. The flower shop has 40 times as many flowers in one cooler as Julia has in her bouquet. The cooler has 120 flowers. How many flowers are in Julia's bouquet?

Lesson 5:   Multiply multiples of 10, 100, and 1,000 by single digits, recognizing patterns.

©2015 Great Minds. eureka-math.org
G4-M3-M4-SE-1.3.1-01.2016

Name _____ Date _____

Represent the following problem by drawing disks in the place value chart.

1. To solve 20 × 40, think

   (2 tens × 4) × 10 = _____

   20 × (4 × 10) = _____

   20 × 40 = _____

hundreds	tens	ones

2. Draw an area model to represent 20 × 40.

   2 tens × 4 tens = _____ _____

3. Draw an area model to represent 30 × 40.

   3 tens × 4 tens = _____ _____

   30 × 40 = _____

EUREKA MATH™

Lesson 6:   Multiply two-digit multiples of 10 by two-digit multiples of 10 with the area model.

©2015 Great Minds. eureka-math.org
G4-M3-M4-SE-1.3.1-01.2016

31

4. Draw an area model to represent 20 × 50.

2 tens × 5 tens = _____ _____

20 × 50 = _____

Rewrite each equation in unit form and solve.

5. 20 × 20 = _____

2 tens × 2 tens = _____ hundreds

6. 60 × 20 = _____

6 tens × 2 _____ = _____ hundreds

7. 70 × 20 = _____

_____ tens × _____ tens = 14 _____

8. 70 × 30 = _____

_____ _____ × _____ _____ = _____ hundreds

Lesson 6:   Multiply two-digit multiples of 10 by two-digit multiples of 10 with the area model.

©2015 Great Minds. eureka-math.org
G4-M3-M4-SE-1.3.1-01.2016

EUREKA MATH™

9. If there are 40 seats per row, how many seats are in 90 rows?

10. One ticket to the symphony costs $50. How much money is collected if 80 tickets are sold?

Lesson 6: Multiply two-digit multiples of 10 by two-digit multiples of 10 with the area model.

©2015 Great Minds. eureka-math.org
G4-M3-M4-SE-1.3.1-01.2016

33

Name _____   Date _____

Represent the following problem by drawing disks in the place value chart.

1. To solve 30 × 60, think

     (3 tens × 6) × 10 = _____

     30 × (6 × 10) = _____

     30 × 60 = _____

hundreds	tens	ones

2. Draw an area model to represent 30 × 60.

     3 tens × 6 tens = _____ _____

3. Draw an area model to represent 20 × 20.

     2 tens × 2 tens = _____ _____

     20 × 20 = _____

     Lesson 6:   Multiply two-digit multiples of 10 by two-digit multiples of 10 with the area model.

©2015 Great Minds. eureka-math.org
G4-M3-M4-SE-1.3.1-01.2016

4. Draw an area model to represent 40 × 60.

4 tens × 6 tens = _____ _____

40 × 60 = _____

Rewrite each equation in unit form and solve.

5. 50 × 20 = _____

5 tens × 2 tens = _____ hundreds

6. 30 × 50 = _____

3 tens × 5 _____ = _____ hundreds

7. 60 × 20 = _____

_____ tens × _____ tens = 12 _____

8. 40 × 70 = _____

____ _____ × ____ _____ = _____ hundreds

Lesson 6:   Multiply two-digit multiples of 10 by two-digit multiples of 10 with the
             area model.

©2015 Great Minds. eureka-math.org
G4-M3-M4-SE-1.3.1-01.2016

35

9. There are 60 seconds in a minute and 60 minutes in an hour. How many seconds are in one hour?

10. To print a comic book, 50 pieces of paper are needed. How many pieces of paper are needed to print 40 comic books?

Lesson 6:   Multiply two-digit multiples of 10 by two-digit multiples of 10 with the area model.

©2015 Great Minds. eureka-math.org

G4-M3-M4-SE-1.3.1-01.2016

Name _____   Date _____

1. Represent the following expressions with disks, regrouping as necessary, writing a matching expression, and recording the partial products vertically as shown below.

a. 1 × 43

tens	ones
● ● ● ●	● ● ●

```
      4  3
  ×      1
  ─────────
         3    → 1 × 3 ones
  +   4  0    → 1 × 4 tens
  ─────────
      4  3
```

b. 2 × 43

tens	ones

c. 3 × 43

hundreds	tens	ones

d. 4 × 43

hundreds	tens	ones

2. Represent the following expressions with disks, regrouping as necessary. To the right, record the partial products vertically.

a. 2 × 36

hundreds	tens	ones

b. 3 × 61

hundreds	tens	ones

c. 4 × 84

hundreds	tens	ones

Lesson 7:   Use place value disks to represent two-digit by one-digit multiplication.

©2015 Great Minds. eureka-math.org
G4-M3-M4-SE-1.3.1-01.2016

Name _____   Date _____

1. Represent the following expressions with disks, regrouping as necessary, writing a matching expression, and recording the partial products vertically.

    a. 3 × 24

tens	ones

    b. 3 × 42

hundreds	tens	ones

    c. 4 × 34

hundreds	tens	ones

EUREKA
MATH™

Lesson 7:   Use place value disks to represent two-digit by one-digit multiplication.

39

©2015 Great Minds. eureka-math.org
G4-M3-M4-SE-1.3.1-01.2016

2. Represent the following expressions with disks, regrouping as necessary. To the right, record the partial products vertically.

   a. 4 × 27

hundreds	tens	ones

   b. 5 × 42

hundreds	tens	ones

3. Cindy says she found a shortcut for doing multiplication problems. When she multiplies 3 × 24, she says, "3 × 4 is 12 ones, or 1 ten and 2 ones. Then, there's just 2 tens left in 24, so add it up, and you get 3 tens and 2 ones." Do you think Cindy's shortcut works? Explain your thinking in words, and justify your response using a model or partial products.

Lesson 7:   Use place value disks to represent two-digit by one-digit multiplication.

EUREKA MATH

©2015 Great Minds. eureka-math.org
G4-M3-M4-SE-1.3.1-01.2016

ten thousands	thousands ,	hundreds	tens	ones

ten thousands place value chart

©2015 Great Minds. eureka-math.org
G4-M3-M4-SE-1.3.1-01.2016

This page intentionally left blank

Name _____ Date _____

1. Represent the following expressions with disks, regrouping as necessary, writing a matching expression, and recording the partial products vertically as shown below.

   a. 1×213

hundreds	tens	ones

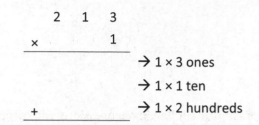

$1 \times$ ___ hundreds $+$ $1 \times$ ___ ten $+$ $1 \times$ ___ ones

   b. 2×213

hundreds	tens	ones

   c. 3×214

hundreds	tens	ones

EUREKA MATH™

Lesson 8: Extend the use of place value disks to represent three- and four-digit by one-digit multiplication.

43

©2015 Great Minds. eureka-math.org
G4-M3-M4-SE-1.3.1-01.2016

d. 3 × 1,254

thousands	hundreds	tens	ones

2. Represent the following expressions with disks, using either method shown during class, regrouping as necessary. To the right, record the partial products vertically.

a. 3 × 212

b. 2 × 4,036

Lesson 8: Extend the use of place value disks to represent three- and four-digit by one-digit multiplication.

©2015 Great Minds. eureka-math.org
G4-M3-M4-SE-1.3.1-01.2016

c.   3 × 2,546

d.   3 × 1,407

3.   Every day at the bagel factory, Cyndi makes 5 different kinds of bagels. If she makes 144 of each kind, what is the total number of bagels that she makes?

Lesson 8:   Extend the use of place value disks to represent three- and four-digit by one-digit multiplication.

©2015 Great Minds. eureka-math.org
G4-M3-M4-SE-1.3.1-01.2016

45

Name _____   Date _____

1. Represent the following expressions with disks, regrouping as necessary, writing a matching expression, and recording the partial products vertically as shown below.

    a.   2 × 424

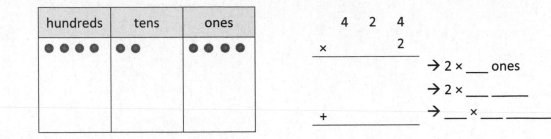

2 × ___ _____ + 2 × ___ _____ + 2 × ___ ones

    b.   3 × 424

hundreds	tens	ones

    c.   4 × 1,424

Lesson 8:   Extend the use of place value disks to represent three- and four-digit
by one-digit multiplication.

©2015 Great Minds. eureka-math.org
G4-M3-M4-SE-1.3.1-01.2016

2. Represent the following expressions with disks, using either method shown in class, regrouping as necessary. To the right, record the partial products vertically.

 a. 2×617

 b. 5×642

 c. $3 \times 3,034$

Lesson 8: Extend the use of place value disks to represent three- and four-digit by one-digit multiplication.

©2015 Great Minds. eureka-math.org
G4-M3-M4-SE-1.3.1-01.2016

47

3. Every day, Penelope jogs three laps around the playground to keep in shape. The playground is rectangular with a width of 163 m and a length of 320 m.

    a.   Find the total amount of meters in one lap.

    b.   Determine how many meters Penelope jogs in three laps.

Lesson 8:   Extend the use of place value disks to represent three- and four-digit by one-digit multiplication.

©2015 Great Minds. eureka-math.org
G4-M3-M4-SE-1.3.1-01.2016

Name _____   Date _____

1. Solve using each method.

Partial Products	Standard Algorithm
a. 3 4 × 4	3 4 × 4

Partial Products	Standard Algorithm
b. 2 2 4 × 3	2 2 4 × 3

2. Solve. Use the standard algorithm.

a. 2 5 1 × 3	b. 1 3 5 × 6	c. 3 0 4 × 9
d. 4 0 5 × 4	e. 3 1 6 × 5	f. 3 9 2 × 6

Lesson 9:   Multiply three- and four-digit numbers by one-digit numbers
applying the standard algorithm.

©2015 Great Minds. eureka-math.org
G4-M3-M4-SE-1.3.1-01.2016

49

3.   The product of 7 and 86 is _____.

4.   9 times as many as 457 is _____.

5.   Jashawn wants to make 5 airplane propellers.
     He needs 18 centimeters of wood for each propeller.
     How many centimeters of wood will he use?

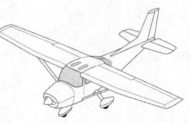

Lesson 9:   Multiply three- and four-digit numbers by one-digit numbers
                               applying the standard algorithm.

©2015 Great Minds. eureka-math.org
G4-M3-M4-SE-1.3.1-01.2016

Name _____   Date _____

1. Solve using each method.

Partial Products	Standard Algorithm

a.   3 4
    ×   4
	3 4
	    × 4

Partial Products	Standard Algorithm

b.   2 2 4
    ×   3
	2 2 4
	    × 3

2. Solve. Use the standard algorithm.

a.   2 5 1
    ×   3
	b. 1 3 5
	    × 6
		c. 3 0 4
		    × 9

d.   4 0 5
    ×   4
	e. 3 1 6
	    × 5
		f. 3 9 2
		    × 6

EUREKA
MATH™

Lesson 9:   Multiply three- and four-digit numbers by one-digit numbers
            applying the standard algorithm.

49

©2015 Great Minds. eureka-math.org
G4-M3-M4-SE-1.3.1-01.2016

3. The product of 7 and 86 is _____.

4. 9 times as many as 457 is _____.

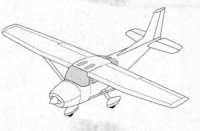

5. Jashawn wants to make 5 airplane propellers.
    He needs 18 centimeters of wood for each propeller.
    How many centimeters of wood will he use?

Lesson 9: Multiply three- and four-digit numbers by one-digit numbers applying the standard algorithm.

©2015 Great Minds. eureka-math.org
G4-M3-M4-SE-1.3.1-01.2016

6. One game system costs $238. How much will 4 game systems cost?

7. A small bag of chips weighs 48 grams. A large bag of chips weighs three times as much as the small bag. How much will 7 large bags of chips weigh?

Lesson 9: Multiply three- and four-digit numbers by one-digit numbers applying the standard algorithm.

©2015 Great Minds. eureka-math.org
G4-M3-M4-SE-1.3.1-01.2016

51

Name _____ Date _____

1. Solve using each method.

Partial Products	Standard Algorithm

a.   4 6
 	
    ×   2
	4 6
	
	    × 2

Partial Products	Standard Algorithm

b.   3 1 5
 	
    ×   4
	3 1 5
	
	    × 4

2. Solve using the standard algorithm.

a.   2   3   2
    ×   4
	b. 1 4 2
	    × 6
		c. 3 1 4
		    × 7

d.   4   4   0
    ×   3
	e. 5 0 7
	    × 8
		f. 3 8 4
		    × 9

Lesson 9:   Multiply three- and four-digit numbers by one-digit numbers
applying the standard algorithm.

©2015 Great Minds. eureka-math.org
G4-M3-M4-SE-1.3.1-01.2016

EUREKA
MATH

2. Represent the following expressions with disks, using either method shown in class, regrouping as necessary. To the right, record the partial products vertically.

   a. 2 × 617

   b. 5 × 642

   c. 3 × 3,034

Lesson 8:   Extend the use of place value disks to represent three- and four-digit by one-digit multiplication.

©2015 Great Minds. eureka-math.org
G4-M3-M4-SE-1.3.1-01.2016

47

3. Every day, Penelope jogs three laps around the playground to keep in shape. The playground is rectangular with a width of 163 m and a length of 320 m.

   a. Find the total amount of meters in one lap.

   b. Determine how many meters Penelope jogs in three laps.

©2015 Great Minds. eureka-math.org
G4-M3-M4-SE-1.3.1-01.2016

JREKA
MATH

3. What is the product of 8 and 54?

4. Isabel earned 350 points while she was playing Blasting Robot. Isabel's mom earned 3 times as many points as Isabel. How many points did Isabel's mom earn?

5. To get enough money to go on a field trip, every student in a club has to raise $53 by selling chocolate bars. There are 9 students in the club. How much money does the club need to raise to go on the field trip?

Lesson 9: Multiply three- and four-digit numbers by one-digit numbers applying the standard algorithm.

©2015 Great Minds. eureka-math.org
G4-M3-M4-SE-1.3.1-01.2016

53

6. Mr. Meyers wants to order 4 tablets for his classroom. Each tablet costs $329. How much will all four tablets cost?

7. Amaya read 64 pages last week. Amaya's older brother, Rogelio, read twice as many pages in the same amount of time. Their big sister, Elianna, is in high school and read 4 times as many pages as Rogelio did. How many pages did Elianna read last week?

Lesson 9: Multiply three- and four-digit numbers by one-digit numbers applying the standard algorithm.

©2015 Great Minds. eureka-math.org
G4-M3-M4-SE-1.3.1-01.2016

Name _____ Date _____

1. Solve using the standard algorithm.

a. 3 × 42	b. 6 × 42
c. 6 × 431	d. 3 × 431
e. 3 × 6,212	f. 3 × 3,106
g. 4 × 4,309	h. 4 × 8,618

Lesson 10: Objective: Multiply three- and four-digit numbers by one-digit numbers applying the standard algorithm.

55

©2015 Great Minds. eureka-math.org
G4-M3-M4-SE-1.3.1-01.2016

2.   There are 365 days in a common year. How many days are in 3 common years?

3.   The length of one side of a square city block is 462 meters. What is the perimeter of the block?

4.   Jake ran 2 miles. Jesse ran 4 times as far. There are 5,280 feet in a mile. How many feet did Jesse run?

Lesson 10:   Objective: Multiply three- and four-digit numbers by one-digit
                                        numbers applying the standard algorithm.

©2015 Great Minds. eureka-math.org
G4-M3-M4-SE-1.3.1-01.2016

Name _____   Date _____

1. Solve using the standard algorithm.

a. 3 × 41	b. 9 × 41
c. 7 × 143	d. 7 × 286
e. 4 × 2,048	f. 4 × 4,096
g. 8 × 4,096	h. 4 × 8,192

EUREKA
MATH™

Lesson 10:   Objective: Multiply three- and four-digit numbers by one-digit
              numbers applying the standard algorithm.

©2015 Great Minds. eureka-math.org
G4-M3-M4-SE-1.3.1-01.2016

57

2. Robert's family brings six gallons of water for the players on the football team. If one gallon of water contains 128 fluid ounces, how many fluid ounces are in six gallons?

3. It takes 687 Earth days for the planet Mars to revolve around the sun once. How many Earth days does it take Mars to revolve around the sun four times?

4. Tammy buys a 4-gigabyte memory card for her camera. Dijonea buys a memory card with twice as much storage as Tammy's. One gigabyte is 1,024 megabytes. How many megabytes of storage does Dijonea have on her memory card?

Lesson 10: Objective: Multiply three- and four-digit numbers by one-digit numbers applying the standard algorithm.

©2015 Great Minds. eureka-math.org
G4-M3-M4-SE-1.3.1-01.2016

Name _____   Date _____

1. Solve the following expressions using the standard algorithm, the partial products method, and the area model.

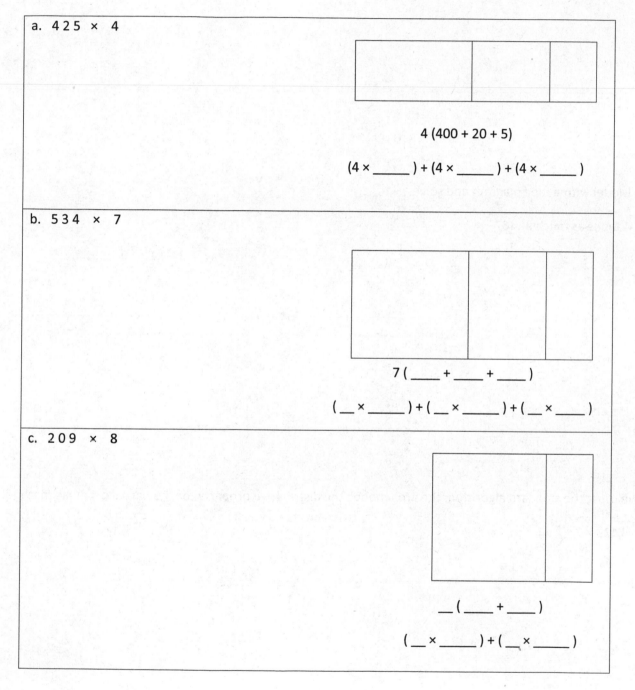

a. 4 2 5 × 4

4 (400 + 20 + 5)

(4 × _____) + (4 × _____) + (4 × _____)

b. 5 3 4 × 7

7 (____ + ____ + ____)

(__ × _____) + (__ × _____) + (__ × _____)

c. 2 0 9 × 8

__ (____ + ____)

(__ × _____) + (__ × _____)

EUREKA
MATH™

Lesson 11:   Connect the area model and the partial products method to the standard algorithm.

59

©2015 Great Minds. eureka-math.org
G4-M3-M4-SE-1.3.1-01.2016

2. Solve using the partial products method.

    Cayla's school has 258 students. Janet's school has 3 times as many students as Cayla's. How many students are in Janet's school?

3. Model with a tape diagram and solve.

    4 times as much as 467

Solve using the standard algorithm, the area model, the distributive property, or the partial products method.

4. 5,131 × 7

©2015 Great Minds. eureka-math.org
G4-M3-M4-SE-1.3.1-01.2016

5. 3 times as many as 2,805

6. A restaurant sells 1,725 pounds of spaghetti and 925 pounds of linguini every month. After 9 months, how many pounds of pasta does the restaurant sell?

Lesson 11:   Connect the area model and the partial products method to the
standard algorithm.

©2015 Great Minds. eureka-math.org
G4-M3-M4-SE-1.3.1-01.2016

61

Name _____   Date _____

1. Solve the following expressions using the standard algorithm, the partial products method, and the area model.

a. 3 0 2 × 8

8 (300 + 2)

(8 × _____) + (8 × _____)

b. 2 1 6 × 5

5 (_____ + _____ + _____)

(__ × _____) + (__ × _____) + (__ × _____)

c. 5 9 3 × 9

__ (_____ + _____ + _____)

(__ × _____) + (__ × _____) + (__ × _____)

Lesson 11:   Connect the area model and the partial products method to the standard algorithm.

©2015 Great Minds. eureka-math.org
G4-M3-M4-SE-1.3.1-01.2016

EUREKA MATH™

2. Solve using the partial products method.

   On Monday, 475 people visited the museum. On Saturday, there were 4 times as many visitors as there were on Monday. How many people visited the museum on Saturday?

3. Model with a tape diagram and solve.

   6 times as much as 384

Solve using the standard algorithm, the area model, the distributive property, or the partial products method.

4. $6,253 \times 3$

Lesson 11: Connect the area model and the partial products method to the standard algorithm.

©2015 Great Minds. eureka-math.org
G4-M3-M4-SE-1.3.1-01.2016

63

5. 7 times as many as 3,073

6. A cafeteria makes 2,516 pounds of white rice and 608 pounds of brown rice every month. After 6 months, how many pounds of rice does the cafeteria make?

Lesson 11: Connect the area model and the partial products method to the standard algorithm.

©2015 Great Minds. eureka-math.org
G4-M3-M4-SE-1.3.1-01.2016

Name _____   Date _____

Use the RDW process to solve the following problems.

1. The table shows the cost of party favors. Each party guest receives a bag with 1 balloon, 1 lollipop, and 1 bracelet. What is the total cost for 9 guests?

Item	Cost
1 balloon	26¢
1 lollipop	14¢
1 bracelet	33¢

2. The Turner family uses 548 liters of water per day. The Hill family uses 3 times as much water per day. How much water does the Hill family use per week?

3. Jayden has 347 marbles. Elvis has 4 times as many as Jayden. Presley has 799 fewer than Elvis. How many marbles does Presley have?

4.   a.   Write an equation that would allow someone to find the value of R.

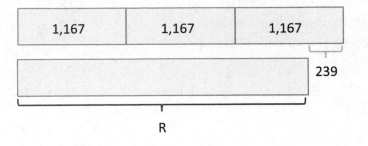

b.   Write your own word problem to correspond to the tape diagram, and then solve.

    Lesson 12:   Solve two-step word problems, including multiplicative comparison.

©2015 Great Minds. eureka-math.org
G4-M3-M4-SE-1.3.1-01.2016

Name _____ Date _____

Use the RDW process to solve the following problems.

1. The table shows the number of stickers of various types in Chrissy's new sticker book. Chrissy's six friends each own the same sticker book. How many stickers do Chrissy and her six friends have altogether?

Type of Sticker	Number of Stickers
flowers	32
smiley faces	21
hearts	39

2. The small copier makes 437 copies each day. The large copier makes 4 times as many copies each day. How many copies does the large copier make each week?

3. Jared sold 194 Boy Scout chocolate bars. Matthew sold three times as many as Jared. Gary sold 297 fewer than Matthew. How many bars did Gary sell?

Lesson 12:   Solve two-step word problems, including multiplicative comparison.

67

©2015 Great Minds. eureka-math.org
G4-M3-M4-SE-1.3.1-01.2016

4. a. Write an equation that would allow someone to find the value of M.

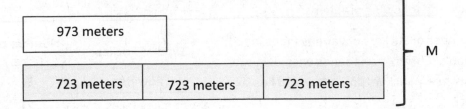

b. Write your own word problem to correspond to the tape diagram, and then solve.

Lesson 12:   Solve two-step word problems, including multiplicative comparison.

©2015 Great Minds. eureka-math.org
G4-M3-M4-SE-1.3.1-01.2016

Name _____   Date _____

Solve using the RDW process.

1. Over the summer, Kate earned $180 each week for 7 weeks. Of that money, she spent $375 on a new computer and $137 on new clothes. How much money did she have left?

2. Sylvia weighed 8 pounds when she was born. By her first birthday, her weight had tripled. By her second birthday, she had gained 12 more pounds. At that time, Sylvia's father weighed 5 times as much as she did. What was Sylvia and her dad's combined weight?

Lesson 13:   Use multiplication, addition, or subtraction to solve multi-step word problems.

©2015 Great Minds. eureka-math.org
G4-M3-M4-SE-1.3.1-01.2016

3. Three boxes weighing 128 pounds each and one box weighing 254 pounds were loaded onto the back of an empty truck. A crate of apples was then loaded onto the same truck. If the total weight loaded onto the truck was 2,000 pounds, how much did the crate of apples weigh?

4. In one month, Charlie read 814 pages. In the same month, his mom read 4 times as many pages as Charlie, and that was 143 pages more than Charlie's dad read. What was the total number of pages read by Charlie and his parents?

Lesson 13:   Use multiplication, addition, or subtraction to solve multi-step word problems.

©2015 Great Minds. eureka-math.org
G4-M3-M4-SE-1.3.1-01.2016

Name _____   Date _____

Solve using the RDW process.

1. A pair of jeans costs $89. A jean jacket costs twice as much. What is the total cost of a jean jacket and 4 pairs of jeans?

2. Sarah bought a shirt on sale for $35. The original price of the shirt was 3 times that amount. Sarah also bought a pair of shoes on sale for $28. The original price of the shoes was 5 times that amount. Together, how much money did the shirt and shoes cost before they went on sale?

Lesson 13:   Use multiplication, addition, or subtraction to solve multi-step word problems.

71

©2015 Great Minds. eureka-math.org
G4-M3-M4-SE-1.3.1-01.2016

3. All 3,000 seats in a theater are being replaced. So far, 5 sections of 136 seats and a sixth section containing 348 seats have been replaced. How many more seats do they still need to replace?

4. Computer Depot sold 762 reams of paper. Paper Palace sold 3 times as much paper as Computer Depot and 143 reams more than Office Supply Central. How many reams of paper were sold by all three stores combined?

Lesson 13:   Use multiplication, addition, or subtraction to solve multi-step word problems.

©2015 Great Minds. eureka-math.org
G4-M3-M4-SE-1.3.1-01.2016

Name _____ Date _____

Use the RDW process to solve the following problems.

1. There are 19 identical socks. How many pairs of socks are there? Will there be any socks without a match? If so, how many?

2. If it takes 8 inches of ribbon to make a bow, how many bows can be made from 3 feet of ribbon (1 foot = 12 inches)? Will any ribbon be left over? If so, how much?

3. The library has 27 chairs and 5 tables. If the same number of chairs is placed at each table, how many chairs can be placed at each table? Will there be any extra chairs? If so, how many?

4. The baker has 42 kilograms of flour. She uses 8 kilograms each day. After how many days will she need to buy more flour?

5. Caleb has 76 apples. He wants to bake as many pies as he can. If it takes 8 apples to make each pie, how many apples will he use? How many apples will not be used?

6. Forty-five people are going to the beach. Seven people can ride in each van. How many vans will be required to get everyone to the beach?

Lesson 14:   Solve division word problems with remainders.

©2015 Great Minds. eureka-math.org
G4-M3-M4-SE-1.3.1-01.2016

Name _____ Date _____

Use the RDW process to solve the following problems.

1. Linda makes booklets using 2 sheets of paper. She has 17 sheets of paper. How many of these booklets can she make? Will she have any extra paper? How many sheets?

2. Linda uses thread to sew the booklets together. She cuts 6 inches of thread for each booklet. How many booklets can she stitch with 50 inches of thread? Will she have any unused thread after stitching up the booklets? If so, how much?

3. Ms. Rochelle wants to put her 29 students into groups of 6. How many groups of 6 can she make? If she puts any remaining students in a smaller group, how many students will be in that group?

4. A trainer gives his horse, Caballo, 7 gallons of water every day from a 57-gallon container. How many days will Caballo receive his full portion of water from the container? On which number day will the trainer need to refill the container of water?

5. Meliza has 43 toy soldiers. She lines them up in rows of 5 to fight imaginary zombies. How many of these rows can she make? After making as many rows of 5 as she can, she puts the remaining soldiers in the last row. How many soldiers are in that row?

6. Seventy-eight students are separated into groups of 8 for a field trip. How many groups are there? The remaining students form a smaller group of how many students?

Lesson 14:   Solve division word problems with remainders.

©2015 Great Minds. eureka-math.org
G4-M3-M4-SE-1.3.1-01.2016

Name _____   Date _____

Show division using an array.	Show division using an area model.

1. 18 ÷ 6

Quotient = _____

Remainder = _____

Can you show 18 ÷ 6 with one rectangle? _____

2. 19 ÷ 6

Quotient = _____

Remainder = _____

Can you show 19 ÷ 6 with one rectangle? _____
Explain how you showed the remainder:

Lesson 15:   Understand and solve division problems with a remainder using the array and area models.

77

©2015 Great Minds. eureka-math.org
G4-M3-M4-SE-1.3.1-01.2016

Solve using an array and an area model. The first one is done for you.

Example: 25 ÷ 2

a.   • • • • • • • • • • • • •
     • • • • • • • • • • • •

     Quotient = 12   Remainder = 1

b.
                        12
        2 ⌷_____⌷|

3.   29 ÷ 3

a.   b.

4.   22 ÷ 5

a.   b.

5.   43 ÷ 4

a.   b.

6.   59 ÷ 7

a.   b.

Lesson 15:   Understand and solve division problems with a remainder using the
                  array and area models.

©2015 Great Minds. eureka-math.org
G4-M3-M4-SE-1.3.1-01.2016

Name _____   Date _____

Show division using an array.	Show division using an area model.
1. 24 ÷ 4 Quotient = _____ Remainder = _____	 Can you show 24 ÷ 4 with one rectangle? _____
2. 25 ÷ 4 Quotient = _____ Remainder = _____	 Can you show 25 ÷ 4 with one rectangle? _____ Explain how you showed the remainder:

Lesson 15:   Understand and solve division problems with a remainder using the array and area models.

79

©2015 Great Minds. eureka-math.org
G4-M3-M4-SE-1.3.1-01.2016

Solve using an array and area model. The first one is done for you.

Example: 25 ÷ 3

a.

Quotient = 8   Remainder = 1

b.

$$
\begin{array}{c}
8 \\
3 \\
\end{array}
$$

3. 44 ÷ 7

a.

b.

4. 34 ÷ 6

a.

b.

5. 37 ÷ 6

a.

b.

6. 46 ÷ 8

a.

b.

Lesson 15:   Understand and solve division problems with a remainder using the array and area models.

©2015 Great Minds. eureka-math.org
G4-M3-M4-SE-1.3.1-01.2016

Name _____   Date _____

Show the division using disks. Relate your work on the place value chart to long division. Check your quotient and remainder by using multiplication and addition.

1. 7 ÷ 2

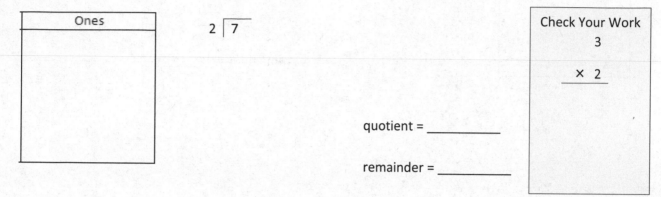

Ones

2 ⟌ 7

quotient = _____

remainder = _____

Check Your Work

3
× 2

2. 27 ÷ 2

Tens	Ones

2 ⟌ 2 7

quotient = _____

remainder = _____

Check Your Work

Lesson 16:   Understand and solve two-digit dividend division problems with a remainder in the ones place by using place value disks.

81

©2015 Great Minds. eureka-math.org
G4-M3-M4-SE-1.3.1-01.2016

3.   8 ÷ 3

Ones

3 ⟌ 8

Check Your Work

quotient = _____

remainder = _____

4.   38 ÷ 3

Tens	Ones

3 ⟌ 3 8

Check Your Work

quotient = _____

remainder = _____

Lesson 16:   Understand and solve two-digit dividend division problems with a remainder in the ones place by using place value disks.

©2015 Great Minds. eureka-math.org
G4-M3-M4-SE-1.3.1-01.2016

EUREKA MATH™

5. 6 ÷ 4

Ones

4 ⟌ 6

quotient = _____

remainder = _____

Check Your Work

6. 86 ÷ 4

Tens	Ones

4 ⟌ 86

quotient = _____

remainder = _____

Check Your Work

Lesson 16: Understand and solve two-digit dividend division problems with a
remainder in the ones place by using place value disks.

83

©2015 Great Minds. eureka-math.org
G4-M3-M4-SE-1.3.1-01.2016

Name _____ Date _____

Show the division using disks. Relate your work on the place value chart to long division. Check your quotient and remainder by using multiplication and addition.

1. 7 ÷ 3

Ones

3 ⟌ 7

quotient = _____

remainder = _____

Check Your Work

2
× 3
_____

2. 67 ÷ 3

Tens	Ones

3 ⟌ 6 7

quotient = _____

remainder = _____

Check Your Work

Lesson 16:   Understand and solve two-digit dividend division problems with a remainder in the ones place by using place value disks.

©2015 Great Minds. eureka-math.org
G4-M3-M4-SE-1.3.1-01.2016

EUREKA MATH

3. 5 ÷ 2

Ones

2 | 5

quotient = _____

remainder = _____

Check Your Work

4. 85 ÷ 2

Tens	Ones

2 | 8 5

quotient = _____

remainder = _____

Check Your Work

Lesson 16:   Understand and solve two-digit dividend division problems with a remainder in the ones place by using place value disks.

85

©2015 Great Minds. eureka-math.org
G4-M3-M4-SE-1.3.1-01.2016

5.   5 ÷ 4

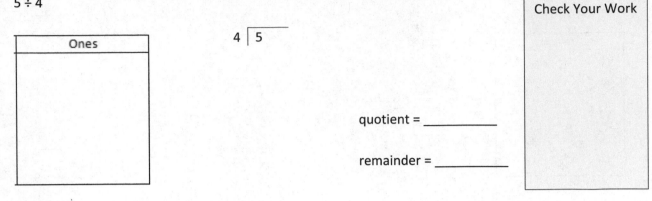

Ones

4 ⟌ 5

Check Your Work

quotient = _____

remainder = _____

6.   85 ÷ 4

Tens	Ones

4 ⟌ 8 5

Check Your Work

quotient = _____

remainder = _____

Lesson 16:   Understand and solve two-digit dividend division problems with a remainder in the ones place by using place value disks.

©2015 Great Minds. eureka-math.org
G4-M3-M4-SE-1.3.1-01.2016

EUREKA MATH™

tens	ones

tens place value chart

Lesson 16:   Understand and solve two-digit dividend division problems with a
remainder in the ones place by using place value disks.

©2015 Great Minds. eureka-math.org
G4-M3-M4-SE-1.3.1-01.2016

87

This page intentionally left blank

Name _____ Date _____

Show the division using disks. Relate your model to long division. Check your quotient and remainder by using multiplication and addition.

1. $5 \div 2$

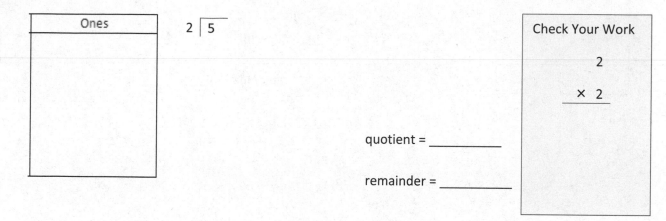

Ones

2 ⟌ 5

quotient = _____

remainder = _____

Check Your Work

2
× 2

2. $50 \div 2$

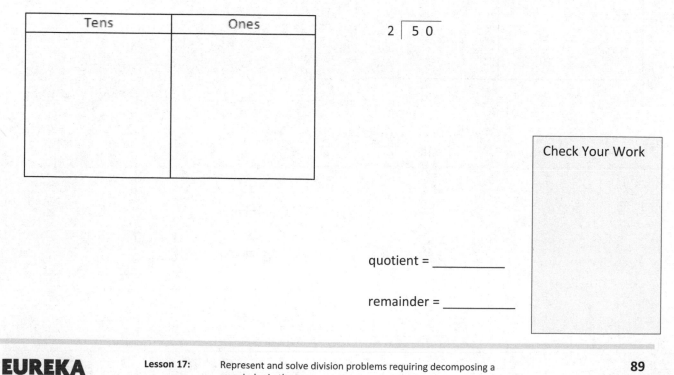

Tens	Ones

2 ⟌ 5 0

quotient = _____

remainder = _____

Check Your Work

Lesson 17: Represent and solve division problems requiring decomposing a remainder in the tens.

89

EUREKA MATH™

©2015 Great Minds. eureka-math.org
G4-M3-M4-SE-1.3.1-01.2016

3. 7 ÷ 3

Ones

3 ⟌ 7

Check Your Work

quotient = _____

remainder = _____

4. 75 ÷ 3

Tens	Ones

3 ⟌ 7 5

Check Your Work

quotient = _____

remainder = _____

Lesson 17: Represent and solve division problems requiring decomposing a remainder in the tens.

©2015 Great Minds. eureka-math.org
G4-M3-M4-SE-1.3.1-01.2016

EUREKA MATH

5. 9 ÷ 4

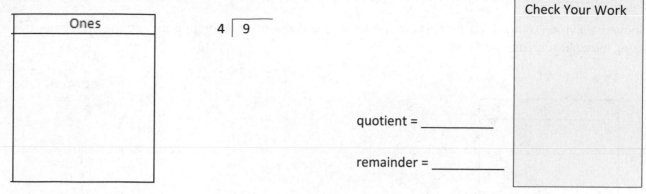

Ones

4 ⟌ 9

quotient = _____

remainder = _____

Check Your Work

6. 92 ÷ 4

Tens	Ones

4 ⟌ 9 2

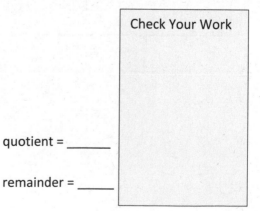

Check Your Work

quotient = _____

remainder = _____

Lesson 17: Represent and solve division problems requiring decomposing a remainder in the tens.

©2015 Great Minds. eureka-math.org
G4-M3-M4-SE-1.3.1-01.2016

91

Name _____   Date _____

Show the division using disks. Relate your model to long division. Check your quotient and remainder by using multiplication and addition.

1. 7 ÷ 2

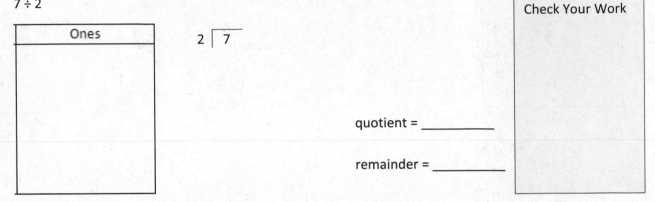

Ones

2 ⟌ 7

Check Your Work

quotient = _____

remainder = _____

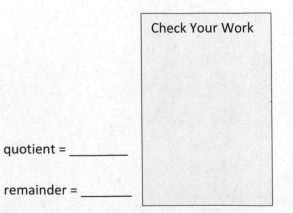

Tens	Ones

2 ⟌ 7 3

Check Your Work

quotient = _____

remainder = _____

Lesson 17:   Represent and solve division problems requiring decomposing a remainder in the tens.

©2015 Great Minds. eureka-math.org
G4-M3-M4-SE-1.3.1-01.2016

3. 6 ÷ 4

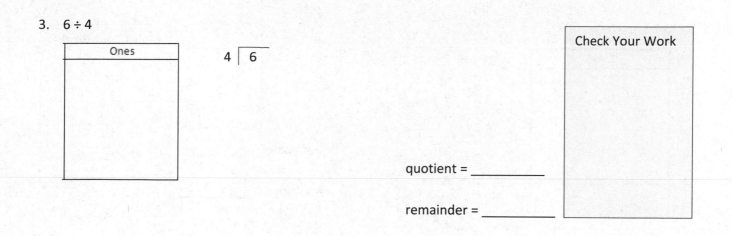

Ones

4 | 6

Check Your Work

quotient = _____

remainder = _____

4. 62 ÷ 4

Tens	Ones

4 | 6 2

Check Your Work

quotient = _____

remainder = _____

Lesson 17:   Represent and solve division problems requiring decomposing a
             remainder in the tens.

93

©2015 Great Minds. eureka-math.org
G4-M3-M4-SE-1.3.1-01.2016

5. 8 ÷ 3

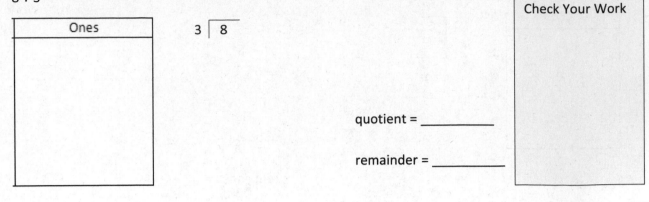

Ones

3 ⟌ 8

Check Your Work

quotient = _____

remainder = _____

6. 84 ÷ 3

Tens	Ones

3 ⟌ 8 4

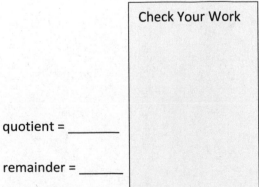

Check Your Work

quotient = _____

remainder = _____

Lesson 17:   Represent and solve division problems requiring decomposing a
remainder in the tens.

©2015 Great Minds. eureka-math.org
G4-M3-M4-SE-1.3.1-01.2016

Name _____   Date _____

Solve using the standard algorithm. Check your quotient and remainder by using multiplication and addition.

1. $46 \div 2$	2. $96 \div 3$
3. $85 \div 5$	4. $52 \div 4$
5. $53 \div 3$	6. $95 \div 4$

©2015 Great Minds. eureka-math.org
G4-M3-M4-SE-1.3.1-01.2016

7. $89 \div 6$	8. $96 \div 6$
9. $60 \div 3$	10. $60 \div 4$
11. $95 \div 8$	12. $95 \div 7$

Lesson 18:   Find whole number quotients and remainders.

EUREKA MATH

©2015 Great Minds. eureka-math.org
G4-M3-M4-SE-1.3.1-01.2016

Name _____ Date _____

Solve using the standard algorithm. Check your quotient and remainder by using multiplication and addition.

1. 84 ÷ 2	2. 84 ÷ 4
3. 48 ÷ 3	4. 80 ÷ 5
5. 79 ÷ 5	6. 91 ÷ 4

©2015 Great Minds. eureka-math.org
G4-M3-M4-SE-1.3.1-01.2016

7.   91 ÷ 6

8.   91 ÷ 7

9.   87 ÷ 3

10.   87 ÷ 6

11.   94 ÷ 8

12.   94 ÷ 6

Lesson 18:   Find whole number quotients and remainders.

©2015 Great Minds. eureka-math.org
G4-M3-M4-SE-1.3.1-01.2016

Name _____   Date _____

1. When you divide 94 by 3, there is a remainder of 1. Model this problem with place value disks. In the place value disk model, how did you show the remainder?

2. Cayman says that 94 ÷ 3 is 30 with a remainder of 4. He reasons this is correct because (3 × 30) + 4 = 94. What mistake has Cayman made? Explain how he can correct his work.

EUREKA
MATH™

Lesson 19:   Explain remainders by using place value understanding and models.

99

©2015 Great Minds. eureka-math.org
G4-M3-M4-SE-1.3.1-01.2016

3. The place value disk model is showing 72 ÷ 3.
    Complete the model. Explain what happens to the
    1 ten that is remaining in the tens column.

4. Two friends evenly share 56 dollars.

    a. They have 5 ten-dollar bills and 6 one-dollar bills. Draw a picture to show how the bills will be
        shared. Will they have to make change at any stage?

    b. Explain how they share the money evenly.

Lesson 19:   Explain remainders by using place value understanding and models.

©2015 Great Minds. eureka-math.org
G4-M3-M4-SE-1.3.1-01.2016

5. Imagine you are filming a video explaining the problem 45 ÷ 3 to new fourth graders. Create a script to explain how you can keep dividing after getting a remainder of 1 ten in the first step.

Lesson 19:   Explain remainders by using place value understanding and models.

©2015 Great Minds. eureka-math.org
G4-M3-M4-SE-1.3.1-01.2016

101

Name _____ Date _____

1. When you divide 86 by 4, there is a remainder of 2. Model this problem with place value disks. In the place value disk model, how can you see that there is a remainder?

2. Francine says that 86 ÷ 4 is 20 with a remainder of 6. She reasons this is correct because (4 × 20) + 6 = 86. What mistake has Francine made? Explain how she can correct her work.

Lesson 19:   Explain remainders by using place value understanding and models.

©2015 Great Minds. eureka-math.org
G4-M3-M4-SE-1.3.1-01.2016

3. The place value disk model is showing 67 ÷ 4. Complete the model. Explain what happens to the 2 tens that are remaining in the tens column.

4. Two friends share 76 blueberries.

   a. To count the blueberries, they put them into small bowls of 10 blueberries. Draw a picture to show how the blueberries can be shared equally. Will they have to split apart any of the bowls of 10 blueberries when they share them?

   b. Explain how the friends can share the blueberries fairly.

Lesson 19:   Explain remainders by using place value understanding and models.

103

©2015 Great Minds. eureka-math.org
G4-M3-M4-SE-1.3.1-01.2016

5. Imagine you are drawing a comic strip showing how to solve the problem 72 ÷ 4 to new fourth graders. Create a script to explain how you can keep dividing after getting a remainder of 3 tens in the first step.

Lesson 19:   Explain remainders by using place value understanding and models.

©2015 Great Minds. eureka-math.org
G4-M3-M4-SE-1.3.1-01.2016

Name _____ Date _____

1. Alfonso solved a division problem by drawing an area model.

    a. Look at the area model. What division problem did Alfonso solve?

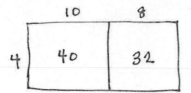

    b. Show a number bond to represent Alfonso's area model. Start with the total, and then show how the total is split into two parts. Below the two parts, represent the total length using the distributive property, and then solve.

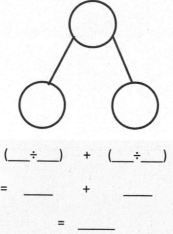

(___÷___) + (___÷___)

= ____ + ____

= ____

2. Solve 45 ÷ 3 using an area model. Draw a number bond, and use the distributive property to solve for the unknown length.

Lesson 20:   Solve division problems without remainders using the area model.

105

©2015 Great Minds. eureka-math.org
G4-M3-M4-SE-1.3.1-01.2016

3. Solve 64 ÷ 4 using an area model. Draw a number bond to show how you partitioned the area, and represent the division with a written method.

4. Solve 92 ÷ 4 using an area model. Explain, using words, pictures, or numbers, the connection of the distributive property to the area model.

5. Solve 72 ÷ 6 using an area model and the standard algorithm.

Lesson 20:   Solve division problems without remainders using the area model.

©2015 Great Minds. eureka-math.org
G4-M3-M4-SE-1.3.1-01.2016

Name _____   Date _____

1. Maria solved a division problem by drawing an area model.

    a. Look at the area model. What division problem did Maria solve?

    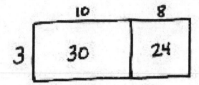

    b. Show a number bond to represent Maria's area model. Start with the total, and then show how the total is split into two parts. Below the two parts, represent the total length using the distributive property, and then solve.

    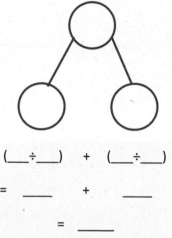

    (___÷___)   +   (___÷___)

    =   _____   +   _____

    =   _____

2. Solve 42 ÷ 3 using an area model. Draw a number bond, and use the distributive property to solve for the unknown length.

EUREKA
MATH™

Lesson 20:   Solve division problems without remainders using the area model.

107

©2015 Great Minds. eureka-math.org
G4-M3-M4-SE-1.3.1-01.2016

3. Solve 60 ÷ 4 using an area model. Draw a number bond to show how you partitioned the area, and represent the division with a written method.

4. Solve 72 ÷ 4 using an area model. Explain, using words, pictures, or numbers, the connection of the distributive property to the area model.

5. Solve 96 ÷ 6 using an area model and the standard algorithm.

Lesson 20:   Solve division problems without remainders using the area model.

©2015 Great Minds. eureka-math.org
G4-M3-M4-SE-1.3.1-01.2016

Name _____ Date _____

1. Solve 37 ÷ 2 using an area model. Use long division and the distributive property to record your work.

2. Solve 76 ÷ 3 using an area model. Use long division and the distributive property to record your work.

3. Carolina solved the following division problem by drawing an area model.

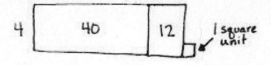

a. What division problem did she solve?

b. Show how Carolina's model can be represented using the distributive property.

Lesson 21: Solve division problems with remainders using the area model.

109

©2015 Great Minds. eureka-math.org
G4-M3-M4-SE-1.3.1-01.2016

Solve the following problems using the area model. Support the area model with long division or the distributive property.

4. 48 ÷ 3	5. 49 ÷ 3
6. 56 ÷ 4	7. 58 ÷ 4
8. 66 ÷ 5	9. 79 ÷ 3

Lesson 21:   Solve division problems with remainders using the area model.

©2015 Great Minds. eureka-math.org
G4-M3-M4-SE-1.3.1-01.2016

10. Seventy-three students are divided into groups of 6 students each. How many groups of 6 students are there? How many students will not be in a group of 6?

Lesson 21:   Solve division problems with remainders using the area model.

111

©2015 Great Minds. eureka-math.org
G4-M3-M4-SE-1.3.1-01.2016

Name _____   Date _____

1. Solve 35 ÷ 2 using an area model. Use long division and the distributive property to record your work.

2. Solve 79 ÷ 3 using an area model. Use long division and the distributive property to record your work.

3. Paulina solved the following division problem by drawing an area model.

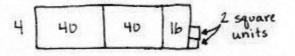

a. What division problem did she solve?

b. Show how Paulina's model can be represented using the distributive property.

©2015 Great Minds. eureka-math.org
G4-M3-M4-SE-1.3.1-01.2016

Solve the following problems using the area model. Support the area model with long division or the distributive property.

4. $42 \div 3$	5. $43 \div 3$
6. $52 \div 4$	7. $54 \div 4$
8. $61 \div 5$	9. $73 \div 3$

©2015 Great Minds. eureka-math.org
G4-M3-M4-SE-1.3.1-01.2016

10. Ninety-seven lunch trays were placed equally in 4 stacks. How many lunch trays were in each stack? How many lunch trays will be left over?

Lesson 21:   Solve division problems with remainders using the area model.

©2015 Great Minds. eureka-math.org
G4-M3-M4-SE-1.3.1-01.2016

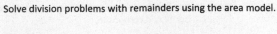

Name _____   Date _____

1. Record the factors of the given numbers as multiplication sentences and as a list in order from least to greatest. Classify each as prime (P) or composite (C). The first problem is done for you.

	Multiplication Sentences	Factors	P or C
a.	4 $1 \times 4 = 4$ $2 \times 2 = 4$	The factors of 4 are: 1, 2, 4	C
b.	6	The factors of 6 are:	
c.	7	The factors of 7 are:	
d.	9	The factors of 9 are:	
e.	12	The factors of 12 are:	
f.	13	The factors of 13 are:	
g.	15	The factors of 15 are:	
h.	16	The factors of 16 are:	
i.	18	The factors of 18 are:	
j.	19	The factors of 19 are:	
k.	21	The factors of 21 are:	
l.	24	The factors of 24 are:	

Lesson 22:   Find factor pairs for numbers to 100, and use understanding of factors
to define prime and composite.

115

©2015 Great Minds. eureka-math.org
G4-M3-M4-SE-1.3.1-01.2016

2. Find all factors for the following numbers, and classify each number as prime or composite. Explain your classification of each as prime or composite.

Factor Pairs for 25	

Factor Pairs for 28	

Factor Pairs for 29	

3. Bryan says all prime numbers are odd numbers.

   a. List all of the prime numbers less than 20 in numerical order.

   b. Use your list to show that Bryan's claim is false.

4. Sheila has 28 stickers to divide evenly among 3 friends. She thinks there will be no leftovers. Use what you know about factor pairs to explain if Sheila is correct.

Lesson 22: Find factor pairs for numbers to 100, and use understanding of factors to define prime and composite.

©2015 Great Minds. eureka-math.org
G4-M3-M4-SE-1.3.1-01.2016

Name _____ Date _____

1. Record the factors of the given numbers as multiplication sentences and as a list in order from least to greatest. Classify each as prime (P) or composite (C). The first problem is done for you.

	Multiplication Sentences	Factors	P or C
a.	8 1 × 4 = 8 2 × 4 = 8	The factors of 8 are: 1, 2, 4, 8	C
b.	10	The factors of 10 are:	
c.	11	The factors of 11 are:	
d.	14	The factors of 14 are:	
e.	17	The factors of 17 are:	
f.	20	The factors of 20 are:	
g.	22	The factors of 22 are:	
h.	23	The factors of 23 are:	
i.	25	The factors of 25 are:	
j.	26	The factors of 26 are:	
k.	27	The factors of 27 are:	
l.	28	The factors of 28 are:	

EUREKA
MATH™

Lesson 22:   Find factor pairs for numbers to 100, and use understanding of factors
             to define prime and composite.

©2015 Great Minds. eureka-math.org
G4-M3-M4-SE-1.3.1-01.2016

117

2. Find all factors for the following numbers, and classify each number as prime or composite. Explain your classification of each as prime or composite.

Factor Pairs for 19	

Factor Pairs for 21	

Factor Pairs for 24	

3. Bryan says that only even numbers are composite.

    a. List all of the odd numbers less than 20 in numerical order.

    b. Use your list to show that Bryan's claim is false.

4. Julie has 27 grapes to divide evenly among 3 friends. She thinks there will be no leftovers. Use what you know about factor pairs to explain whether or not Julie is correct.

Lesson 22:   Find factor pairs for numbers to 100, and use understanding of factors to define prime and composite.

©2015 Great Minds. eureka-math.org
G4-M3-M4-SE-1.3.1-01.2016

Name _____ Date _____

1. Explain your thinking or use division to answer the following.

a. Is 2 a factor of 84?	b. Is 2 a factor of 83?
c. Is 3 a factor of 84?	d. Is 2 a factor of 92?
e. Is 6 a factor of 84?	f. Is 4 a factor of 92?
g. Is 5 a factor of 84?	h. Is 8 a factor of 92?

Lesson 23:   Use division and the associative property to test for factors and
              observe patterns.

©2015 Great Minds. eureka-math.org
G4-M3-M4-SE-1.3.1-01.2016

119

2. Use the associative property to find more factors of 24 and 36.

a. $24 = 12 \times 2$

$ = (\underline{} \times 3) \times 2$

$ = \underline{} \times (3 \times 2)$

$ = \underline{} \times 6$

$ = \underline{}$

b. $36 = \underline{} \times 4$

$ = (\underline{} \times 3) \times 4$

$ = \underline{} \times (3 \times 4)$

$ = \underline{} \times 12$

$ = \underline{}$

3. In class, we used the associative property to show that when 6 is a factor, then 2 and 3 are factors, because $6 = 2 \times 3$. Use the fact that $8 = 4 \times 2$ to show that 2 and 4 are factors of 56, 72, and 80.

$$56 = 8 \times 7 \qquad\qquad 72 = 8 \times 9 \qquad\qquad 80 = 8 \times 10$$

4. The first statement is false. The second statement is true. Explain why, using words, pictures, or numbers.

        If a number has 2 and 4 as factors, then it has 8 as a factor.

        If a number has 8 as a factor, then both 2 and 4 are factors.

Lesson 23:   Use division and the associative property to test for factors and observe patterns.

©2015 Great Minds. eureka-math.org
G4-M3-M4-SE-1.3.1-01.2016

Name _____ Date _____

1. Explain your thinking or use division to answer the following.

a. Is 2 a factor of 72?	b. Is 2 a factor of 73?
c. Is 3 a factor of 72?	d. Is 2 a factor of 60?
e. Is 6 a factor of 72?	f. Is 4 a factor of 60?
g. Is 5 a factor of 72?	h. Is 8 a factor of 60?

Lesson 23:   Use division and the associative property to test for factors and
            observe patterns.

©2015 Great Minds. eureka-math.org
G4-M3-M4-SE-1.3.1-01.2016

121

2. Use the associative property to find more factors of 12 and 30.

a.   12 = 6 × 2   b.   30 = ____ ×   5

        = (___ × 2) × 2   = (____ × 3) × 5

        = ___ × (2 × 2)   = ____ × (3 × 5)

        = ___ × ___   = ____ × 15

        = ___   = ____

3. In class, we used the associative property to show that when 6 is a factor, then 2 and 3 are factors, because 6 = 2 × 3. Use the fact that 10 = 5 × 2 to show that 2 and 5 are factors of 70, 80, and 90.

        70 = 10 × 7   80 = 10 × 8   90 = 10 × 9

4. The first statement is false. The second statement is true. Explain why, using words, pictures, or numbers.

        If a number has 2 and 6 as factors, then it has 12 as a factor.
        If a number has 12 as a factor, then both 2 and 6 are factors.

Lesson 23:   Use division and the associative property to test for factors and
                        observe patterns.

©2015 Great Minds. eureka-math.org
G4-M3-M4-SE-1.3.1-01.2016

Name _____ Date _____

1. For each of the following, time yourself for 1 minute. See how many multiples you can write.

    a. Write the multiples of 5 starting from 100.

    b. Write the multiples of 4 starting from 20.

    c. Write the multiples of 6 starting from 36.

2. List the numbers that have 24 as a multiple.

3. Use mental math, division, or the associative property to solve. (Use scratch paper if you like.)

    a. Is 12 a multiple of 4? _____   Is 4 a factor of 12? _____

    b. Is 42 a multiple of 8? _____   Is 8 a factor of 42? _____

    c. Is 84 a multiple of 6? _____   Is 6 a factor of 84? _____

4. Can a prime number be a multiple of any other number except itself? Explain why or why not.

Lesson 24:   Determine if a whole number is a multiple of another number.

123

©2015 Great Minds. eureka-math.org
G4-M3-M4-SE-1.3.1-01.2016

5. Follow the directions below.

1	2	3	4	5	6	7	8	9	10
11	12	13	14	15	16	17	18	19	20
21	22	23	24	25	26	27	28	29	30
31	32	33	34	35	36	37	38	39	40
41	42	43	44	45	46	47	48	49	50
51	52	53	54	55	56	57	58	59	60
61	62	63	64	65	66	67	68	69	70
71	72	73	74	75	76	77	78	79	80
81	82	83	84	85	86	87	88	89	90
91	92	93	94	95	96	97	98	99	100

a. Circle in red the multiples of 2. When a number is a multiple of 2, what are the possible values for the ones digit?

b. Shade in green the multiples of 3. Choose one. What do you notice about the sum of the digits? Choose another. What do you notice about the sum of the digits?

c. Circle in blue the multiples of 5. When a number is a multiple of 5, what are the possible values for the ones digit?

d. Draw an X over the multiples of 10. What digit do all multiples of 10 have in common?

Lesson 24:   Determine if a whole number is a multiple of another number.

©2015 Great Minds. eureka-math.org
G4-M3-M4-SE-1.3.1-01.2016

Name _____ Date _____

1. For each of the following, time yourself for 1 minute. See how many multiples you can write.

   a. Write the multiples of 5 starting from 75.

   b. Write the multiples of 4 starting from 40.

   c. Write the multiples of 6 starting from 24.

2. List the numbers that have 30 as a multiple.

3. Use mental math, division, or the associative property to solve. (Use scratch paper if you like.)

   a. Is 12 a multiple of 3? _____   Is 3 a factor of 12? _____

   b. Is 48 a multiple of 8? _____   Is 48 a factor of 8? _____

   c. Is 56 a multiple of 6? _____   Is 6 a factor of 56? _____

4. Can a prime number be a multiple of any other number except itself? Explain why or why not.

5. Follow the directions below.

1	2	3	4	5	6	7	8	9	10
11	12	13	14	15	16	17	18	19	20
21	22	23	24	25	26	27	28	29	30
31	32	33	34	35	36	37	38	39	40
41	42	43	44	45	46	47	48	49	50
51	52	53	54	55	56	57	58	59	60
61	62	63	64	65	66	67	68	69	70
71	72	73	74	75	76	77	78	79	80
81	82	83	84	85	86	87	88	89	90
91	92	93	94	95	96	97	98	99	100

a. Underline the multiples of 6. When a number is a multiple of 6, what are the possible values for the ones digit?

b. Draw a square around the multiples of 4. Look at the multiples of 4 that have an odd number in the tens place. What values do they have in the ones place?

c. Look at the multiples of 4 that have an even number in the tens place. What values do they have in the ones place? Do you think this pattern would continue with multiples of 4 that are larger than 100?

d. Circle the multiples of 9. Choose one. What do you notice about the sum of the digits?
Choose another one. What do you notice about the sum of the digits?

Lesson 24:   Determine if a whole number is a multiple of another number.

©2015 Great Minds. eureka-math.org
G4-M3-M4-SE-1.3.1-01.2016

Name _____ Date _____

1. Follow the directions.

   Shade the number 1 red.

   a. Circle the first unmarked number.

   b. Cross off every multiple of that number except the one you circled. If it's already crossed off, skip it.

   c. Repeat Steps (a) and (b) until every number is either circled or crossed off.

   d. Shade every crossed out number in orange.

1	2	3	4	5	6	7	8	9	10
11	12	13	14	15	16	17	18	19	20
21	22	23	24	25	26	27	28	29	30
31	32	33	34	35	36	37	38	39	40
41	42	43	44	45	46	47	48	49	50
51	52	53	54	55	56	57	58	59	60
61	62	63	64	65	66	67	68	69	70
71	72	73	74	75	76	77	78	79	80
81	82	83	84	85	86	87	88	89	90
91	92	93	94	95	96	97	98	99	100

Lesson 25: Explore properties of prime and composite numbers to 100 by using multiples.

127

©2015 Great Minds. eureka-math.org
G4-M3-M4-SE-1.3.1-01.2016

2. a. List the circled numbers.

b. Why were the circled numbers not crossed off along the way?

c. Except for the number 1, what is similar about all of the numbers that were crossed off?

d. What is similar about all of the numbers that were circled?

Lesson 25: Explore properties of prime and composite numbers to 100 by using multiples.

©2015 Great Minds. eureka-math.org
G4-M3-M4-SE-1.3.1-01.2016

Name _____   Date _____

1. A student used the sieve of Eratosthenes to find all prime numbers less than 100. Create a step-by-step set of directions to show how it was completed. Use the word bank to help guide your thinking as you write the directions. Some words may be used just once, more than once, or not at all.

Word Bank	
composite	cross out
number	shade
circle	X
multiple	prime

Directions for completing the sieve of Eratosthenes activity:

Lesson 25: Explore properties of prime and composite numbers to 100 by using multiples.

©2015 Great Minds. eureka-math.org
G4-M3-M4-SE-1.3.1-01.2016

129

2. What do all of the numbers that are crossed out have in common?

3. What do all of the circled numbers have in common?

4. There is one number that is neither crossed out nor circled. Why is it treated differently?

Lesson 25:   Explore properties of prime and composite numbers to 100 by using
              multiples.

©2015 Great Minds. eureka-math.org
G4-M3-M4-SE-1.3.1-01.2016

Name _____ Date _____

1. Draw place value disks to represent the following problems. Rewrite each in unit form and solve.

   a. 6 ÷ 2 = _____

      ① ① ①   ① ① ①

      6 ones ÷ 2 = _____ ones

   b. 60 ÷ 2 = _____

      6 tens ÷ 2 = _____

   c. 600 ÷ 2 = _____

      _____ ÷ 2 = _____

   d. 6,000 ÷ 2 = _____

      _____ ÷ 2 = _____

2. Draw place value disks to represent each problem. Rewrite each in unit form and solve.

   a. 12 ÷ 3 = _____

      12 ones ÷ 3 = _____ ones

   b. 120 ÷ 3 = _____

      _____ ÷ 3 = _____

   c. 1,200 ÷ 3 = _____

      _____ ÷ 3 = _____

EUREKA MATH™

Lesson 26:   Divide multiples of 10, 100, and 1,000 by single-digit numbers.

131

©2015 Great Minds. eureka-math.org
G4-M3-M4-SE-1.3.1-01.2016

3. Solve for the quotient. Rewrite each in unit form.

a. 800 ÷ 2 = 400 8 hundreds ÷ 2 = 4 hundreds	b. 600 ÷ 2 = _____	c. 800 ÷ 4 = _____	d. 900 ÷ 3 = _____
e. 300 ÷ 6 = _____ 30 tens ÷ 6 = ____ tens	f. 240 ÷ 4 = _____	g. 450 ÷ 5 = _____	h. 200 ÷ 5 = _____
i. 3,600 ÷ 4 = _____ 36 hundreds ÷ 4 = _____ hundreds	j. 2,400 ÷ 4 = _____	k. 2,400 ÷ 3 = _____	l. 4,000 ÷ 5 = _____

4. Some sand weighs 2,800 kilograms. It is divided equally among 4 trucks. How many kilograms of sand are in each truck?

©2015 Great Minds. eureka-math.org
G4-M3-M4-SE-1.3.1-01.2016

5. Ivy has 5 times as many stickers as Adrian has. Ivy has 350 stickers. How many stickers does Adrian have?

6. An ice cream stand sold $1,600 worth of ice cream on Saturday, which was 4 times the amount sold on Friday. How much money did the ice cream stand collect on Friday?

Name _____   Date _____

1. Draw place value disks to represent the following problems. Rewrite each in unit form and solve.

   a. 6 ÷ 3 = _____   ①① ①① ①①

       6 ones ÷ 3 = _____ones

   b. 60 ÷ 3 = _____

       6 tens ÷ 3 = _____

   c. 600 ÷ 3 = _____

       _____ ÷ 3 = _____

   d. 6,000 ÷ 3 = _____

       _____ ÷ 3 = _____

2. Draw place value disks to represent each problem. Rewrite each in unit form and solve.

   a. 12 ÷ 4 = _____

       12 ones ÷ 4 = _____ones

   b. 120 ÷ 4 = _____

       _____ ÷ 4 = _____

   c. 1,200 ÷ 4 = _____

       _____ ÷ 4 = _____

Lesson 26:   Divide multiples of 10, 100, and 1,000 by single-digit numbers.

©2015 Great Minds. eureka-math.org
G4-M3-M4-SE-1.3.1-01.2016

EUREKA MATH

3. Solve for the quotient. Rewrite each in unit form.

a. $800 \div 4 = 200$ 8 hundreds ÷ 4 = 2 hundreds	b. $900 \div 3 = $ _____	c. $400 \div 2 = $ _____	d. $300 \div 3 = $ _____
e. $200 \div 4 = $ _____ 20 tens ÷ 4 = ____ tens	f. $160 \div 2 = $ _____	g. $400 \div 5 = $ _____	h. $300 \div 5 = $ _____
i. $1,200 \div 3 = $ _____ 12 hundreds ÷ 3 = ____ hundreds	j. $1,600 \div 4 = $ _____	k. $2,400 \div 4 = $ _____	l. $3,000 \div 5 = $ _____

4. A fleet of 5 fire engines carries a total of 20,000 liters of water. If each truck holds the same amount of water, how many liters of water does each truck carry?

©2015 Great Minds. eureka-math.org
G4-M3-M4-SE-1.3.1-01.2016

5. Jamie drank 4 times as much juice as Brodie. Jamie drank 280 milliliters of juice. How much juice did Brodie drink?

6. A diner sold $2,400 worth of French fries in June, which was 4 times as much as was sold in May. How many dollars' worth of French fries were sold at the diner in May?

EUREKA
MATH™

©2015 Great Minds. eureka-math.org
G4-M3-M4-SE-1.3.1-01.2016

ones	
tens	
hundreds	
thousands	

thousands place value chart for dividing

Lesson 26:   Divide multiples of 10, 100, and 1,000 by single-digit numbers.

137

©2015 Great Minds. eureka-math.org
G4-M3-M4-SE-1.3.1-01.2016

This page intentionally left blank

Name _____ Date _____

1. Divide. Use place value disks to model each problem.

a. 324 ÷ 2

b. 344 ÷ 2

Lesson 27:   Represent and solve division problems with up to a three-digit
dividend numerically and with place value disks requiring
decomposing a remainder in the hundreds place.

139

©2015 Great Minds. eureka-math.org
G4-M3-M4-SE-1.3.1-01.2016

EUREKA MATH™

c. 483 ÷ 3

d. 549 ÷ 3

Represent and solve division problems with up to a three-digit
dividend numerically and with place value disks requiring
decomposing a remainder in the hundreds place.

©2015 Great Minds. eureka-math.org
G4-M3-M4-SE-1.3.1-01.2016

2. Model using place value disks and record using the algorithm.

a. 655 ÷ 5
    Disks   Algorithm

b. 726÷ 3
    Disks   Algorithm

c. 688 ÷ 4
    Disks   Algorithm

Lesson 27:   Represent and solve division problems with up to a three-digit
             dividend numerically and with place value disks requiring
             decomposing a remainder in the hundreds place.

©2015 Great Minds. eureka-math.org
G4-M3-M4-SE-1.3.1-01.2016

141

Name _____   Date _____

1. Divide. Use place value disks to model each problem.

a. $346 \div 2$

b. $528 \div 2$

Lesson 27:   Represent and solve division problems with up to a three-digit dividend numerically and with place value disks requiring decomposing a remainder in the hundreds place.

©2015 Great Minds. eureka-math.org
G4-M3-M4-SE-1.3.1-01.2016

c. 516 ÷ 3

d. 729 ÷ 3

Lesson 27: Represent and solve division problems with up to a three-digit dividend numerically and with place value disks requiring decomposing a remainder in the hundreds place.

143

©2015 Great Minds. eureka-math.org
G4-M3-M4-SE-1.3.1-01.2016

2. Model using place value disks, and record using the algorithm.

a. 648 ÷ 4
    Disks   Algorithm

b. 755 ÷ 5
    Disks   Algorithm

c. 964 ÷ 4
    Disks   Algorithm

Lesson 27:   Represent and solve division problems with up to a three-digit
                  dividend numerically and with place value disks requiring
                  decomposing a remainder in the hundreds place.

©2015 Great Minds. eureka-math.org
G4-M3-M4-SE-1.3.1-01.2016

Name _____   Date _____

1. Divide. Check your work by multiplying. Draw disks on a place value chart as needed.

---

a. 574 ÷ 2

---

b. 861 ÷ 3

---

c. 354 ÷ 2

Lesson 28:   Represent and solve three-digit dividend division with divisors
of 2, 3, 4, and 5 numerically.

©2015 Great Minds. eureka-math.org
G4-M3-M4-SE-1.3.1-01.2016

145

d. 354 ÷ 3

e. 873 ÷ 4

f. 591 ÷ 5

Lesson 28:   Represent and solve three-digit dividend division with divisors
of 2, 3, 4, and 5 numerically.

©2015 Great Minds. eureka-math.org
G4-M3-M4-SE-1.3.1-01.2016

g. 275 ÷ 3

h. 459 ÷ 5

i. 678 ÷ 4

Lesson 28:   Represent and solve three-digit dividend division with divisors
of 2, 3, 4, and 5 numerically.

147

©2015 Great Minds. eureka-math.org
G4-M3-M4-SE-1.3.1-01.2016

j.   955 ÷ 4

2. Zach filled 581 one-liter bottles with apple cider. He distributed the bottles to 4 stores. Each store received the same number of bottles. How many liter bottles did each of the stores receive? Were there any bottles left over? If so, how many?

Lesson 28:   Represent and solve three-digit dividend division with divisors of 2, 3, 4, and 5 numerically.

©2015 Great Minds. eureka-math.org
G4-M3-M4-SE-1.3.1-01.2016

Name _____   Date _____

1. Divide. Check your work by multiplying. Draw disks on a place value chart as needed.

a.   378 ÷ 2

b.   795 ÷ 3

c.   512 ÷ 4

Lesson 28:   Represent and solve three-digit dividend division with divisors
of 2, 3, 4, and 5 numerically.

©2015 Great Minds. eureka-math.org
G4-M3-M4-SE-1.3.1-01.2016

149

d. 492 ÷ 4

e. 539 ÷ 3

f. 862 ÷ 5

Lesson 28:   Represent and solve three-digit dividend division with divisors
              of 2, 3, 4, and 5 numerically.

©2015 Great Minds. eureka-math.org
G4-M3-M4-SE-1.3.1-01.2016

EUREKA
MATH

g. 498 ÷ 3

h. 783 ÷ 5

i. 621 ÷ 4

Lesson 28: Represent and solve three-digit dividend division with divisors of 2, 3, 4, and 5 numerically.

151

©2015 Great Minds. eureka-math.org
G4-M3-M4-SE-1.3.1-01.2016

j. 531 ÷ 4

2. Selena's dog completed an obstacle course that was 932 meters long. There were 4 parts to the course, all equal in length. How long was 1 part of the course?

Represent and solve three-digit dividend division with divisors of 2, 3, 4, and 5 numerically.

©2015 Great Minds. eureka-math.org
G4-M3-M4-SE-1.3.1-01.2016

Name _____ Date _____

1. Divide, and then check using multiplication.

a. 1,672 ÷ 4

b. 1,578 ÷ 4

c. 6,948 ÷ 2

Lesson 29: Represent numerically four-digit dividend division with divisors
of 2, 3, 4, and 5, decomposing a remainder up to three times.

©2015 Great Minds. eureka-math.org
G4-M3-M4-SE-1.3.1-01.2016

153

d.   8,949 ÷ 4

e.   7,569 ÷ 2

f.   7,569 ÷ 3

Lesson 29:   Represent numerically four-digit dividend division with divisors of 2, 3, 4, and 5, decomposing a remainder up to three times.

©2015 Great Minds. eureka-math.org
G4-M3-M4-SE-1.3.1-01.2016

EUREKA
MATH

g. 7,955 ÷ 5

h. 7,574 ÷ 5

i. 7,469 ÷ 3

Lesson 29: Represent numerically four-digit dividend division with divisors of 2, 3, 4, and 5, decomposing a remainder up to three times.

©2015 Great Minds. eureka-math.org
G4-M3-M4-SE-1.3.1-01.2016

155

j. 9,956 ÷ 4

2. There are twice as many cows as goats on a farm. All the cows and goats have a total of 1,116 legs.
How many goats are there?

Lesson 29:   Represent numerically four-digit dividend division with divisors
of 2, 3, 4, and 5, decomposing a remainder up to three times.

©2015 Great Minds. eureka-math.org
G4-M3-M4-SE-1.3.1-01.2016

EUREKA
MATH™

Name _____   Date _____

1. Divide, and then check using multiplication.

    a. 2,464 ÷ 4

    b. 1,848 ÷ 3

    c. 9,426 ÷ 3

Lesson 29:   Represent numerically four-digit dividend division with divisors
of 2, 3, 4, and 5, decomposing a remainder up to three times.

©2015 Great Minds. eureka-math.org
G4-M3-M4-SE-1.3.1-01.2016

157

d. 6,587 ÷ 2

e. 5,445 ÷ 3

f. 5,425 ÷ 2

Lesson 29: Represent numerically four-digit dividend division with divisors of 2, 3, 4, and 5, decomposing a remainder up to three times.

©2015 Great Minds. eureka-math.org
G4-M3-M4-SE-1.3.1-01.2016

g.   8,467 ÷ 3

h.   8,456 ÷ 3

i.   4,937 ÷ 4

Lesson 29:   Represent numerically four-digit dividend division with divisors
             of 2, 3, 4, and 5, decomposing a remainder up to three times.

159

©2015 Great Minds. eureka-math.org
G4-M3-M4-SE-1.3.1-01.2016

j. $6,173 \div 5$

2.   A truck has 4 crates of apples. Each crate has an equal number of apples. Altogether, the truck is carrying 1,728 apples. How many apples are in 3 crates?

Lesson 29:   Represent numerically four-digit dividend division with divisors
of 2, 3, 4, and 5, decomposing a remainder up to three times.

©2015 Great Minds. eureka-math.org
G4-M3-M4-SE-1.3.1-01.2016

EUREKA
MATH

Name _____   Date _____

Divide. Check your solutions by multiplying.

1. 204 ÷ 4

2. 704 ÷ 3

3. 627 ÷ 3

4. 407 ÷ 2

Lesson 30:   Solve division problems with a zero in the dividend or with a zero in
             the quotient.

©2015 Great Minds. eureka-math.org
G4-M3-M4-SE-1.3.1-01.2016

161

5.   760 ÷ 4

6.   5,120 ÷ 4

7.   3,070 ÷ 5

8.   6,706 ÷ 5

Lesson 30:   Solve division problems with a zero in the dividend or with a zero in
                        the quotient.

©2015 Great Minds. eureka-math.org
G4-M3-M4-SE-1.3.1-01.2016

9. 8,313 ÷ 4

10. 9,008 ÷ 3

11. a. Find the quotient and remainder for 3,131 ÷ 3.

    b. How could you change the digit in the ones place of the whole so that there would be no remainder? Explain how you determined your answer.

Lesson 30:   Solve division problems with a zero in the dividend or with a zero in the quotient.

©2015 Great Minds. eureka-math.org
G4-M3-M4-SE-1.3.1-01.2016

163

Name _____ Date _____

Divide. Check your solutions by multiplying.

1. 409 ÷ 5

2. 503 ÷ 2

3. 831 ÷ 4

4. 602 ÷ 3

Lesson 30:   Solve division problems with a zero in the dividend or with a zero in the quotient.

©2015 Great Minds. eureka-math.org
G4-M3-M4-SE-1.3.1-01.2016

5. 720 ÷ 3

6. 6,250 ÷ 5

7. 2,060 ÷ 5

8. 9,031 ÷ 2

Lesson 30: Solve division problems with a zero in the dividend or with a zero in the quotient.

165

©2015 Great Minds. eureka-math.org
G4-M3-M4-SE-1.3.1-01.2016

9.   6,218 ÷ 4

10.   8,000 ÷ 4

Lesson 30:   Solve division problems with a zero in the dividend or with a zero in the quotient.

©2015 Great Minds. eureka-math.org
G4-M3-M4-SE-1.3.1-01.2016

Name _____ Date _____

Draw a tape diagram and solve. The first two tape diagrams have been drawn for you. Identify if the group size or the number of groups is unknown.

1. Monique needs exactly 4 plates on each table for the banquet. If she has 312 plates, how many tables is she able to prepare?

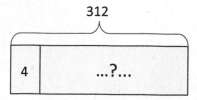

2. 2,365 books were donated to an elementary school. If 5 classrooms shared the books equally, how many books did each class receive?

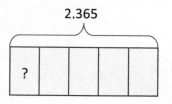

3. If 1,503 kilograms of rice was packed in sacks weighing 3 kilograms each, how many sacks were packed?

4. Rita made 5 batches of cookies. There was a total of 2,400 cookies. If each batch contained the same number of cookies, how many cookies were in 4 batches?

5. Every day, Sarah drives the same distance to work and back home. If Sarah drove 1,005 miles in 5 days, how far did Sarah drive in 3 days?

Name _____   Date _____

Solve the following problems. Draw tape diagrams to help you solve. Identify if the group size or the number of groups is unknown.

1. 500 milliliters of juice was shared equally by 4 children. How many milliliters of juice did each child get?

2. Kelly separated 618 cookies into baggies. Each baggie contained 3 cookies. How many baggies of cookies did Kelly make?

3. Jeff biked the same distance each day for 5 days. If he traveled 350 miles altogether, how many miles did he travel each day?

EUREKA
MATH™

Lesson 31:   Interpret division word problems as either *number of groups unknown*
             or *group size unknown*

©2015 Great Minds. eureka-math.org
G4-M3-M4-SE-1.3.1-01.2016

169

4. A piece of ribbon 876 inches long was cut by a machine into 4-inch long strips to be made into bows. How many strips were cut?

5. Five Martians equally share 1,940 Groblarx fruits. How many Groblarx fruits will 3 of the Martians receive?

Lesson 31:   Interpret division word problems as either *number of groups unknown* or *group size unknown*

©2015 Great Minds. eureka-math.org
G4-M3-M4-SE-1.3.1-01.2016

Name _____   Date _____

Solve the following problems. Draw tape diagrams to help you solve. If there is a remainder, shade in a small portion of the tape diagram to represent that portion of the whole.

1. A concert hall contains 8 sections of seats with the same number of seats in each section. If there are 248 seats, how many seats are in each section?

2. In one day, the bakery made 719 bagels. The bagels were divided into 9 equal shipments. A few bagels were left over and given to the baker. How many bagels did the baker get?

3. The sweet shop has 614 pieces of candy. They packed the candy into bags with 7 pieces in each bag. How many bags of candy did they fill? How many pieces of candy were left?

EUREKA MATH

Lesson 32:   Interpret and find whole number quotients and remainders to solve one-step division word problems with larger divisors of 6, 7, 8, and 9.

©2015 Great Minds. eureka-math.org
G4-M3-M4-SE-1.3.1-01.2016

171

4. There were 904 children signed up for the relay race. If there were 6 children on each team, how many teams were made? The remaining children served as referees. How many children served as referees?

5. 1,188 kilograms of rice are divided into 7 sacks. How many kilograms of rice are in 6 sacks of rice? How many kilograms of rice remain?

Lesson 32:   Interpret and find whole number quotients and remainders to solve one-step division word problems with larger divisors of 6, 7, 8, and 9.

©2015 Great Minds. eureka-math.org
G4-M3-M4-SE-1.3.1-01.2016

Name _____ Date _____

Solve the following problems. Draw tape diagrams to help you solve. If there is a remainder, shade in a small portion of the tape diagram to represent that portion of the whole.

1. Meneca bought a package of 435 party favors to give to the guests at her birthday party. She calculated that she could give 9 party favors to each guest. How many guests is she expecting?

2. 4,000 pencils were donated to an elementary school. If 8 classrooms shared the pencils equally, how many pencils did each class receive?

3. 2,008 kilograms of potatoes were packed into sacks weighing 8 kilograms each. How many sacks were packed?

**EUREKA
MATH**™

Lesson 32:   Interpret and find whole number quotients and remainders to solve one-step division word problems with larger divisors of 6, 7, 8, and 9.

173

©2015 Great Minds. eureka-math.org
G4-M3-M4-SE-1.3.1-01.2016

4. A baker made 7 batches of muffins. There was a total of 252 muffins. If there was the same number of muffins in each batch, how many muffins were in a batch?

5. Samantha ran 3,003 meters in 7 days. If she ran the same distance each day, how far did Samantha run in 3 days?

Lesson 32:   Interpret and find whole number quotients and remainders to solve one-step division word problems with larger divisors of 6, 7, 8, and 9.

©2015 Great Minds. eureka-math.org
G4-M3-M4-SE-1.3.1-01.2016

Name _____ Date _____

1. Ursula solved the following division problem by drawing an area model.

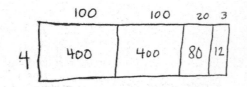

a. What division problem did she solve?

b. Show a number bond to represent Ursula's area model, and represent the total length using the distributive property.

2. a. Solve 960 ÷ 4 using the area model. There is no remainder in this problem.

b. Draw a number bond and use the long division algorithm to record your work from Part (a).

Lesson 33:   Explain the connection of the area model of division to the long
              division algorithm for three and four digit dividends.

©2015 Great Minds. eureka-math.org
G4-M3-M4-SE-1.3.1-01.2016

175

3. a. Draw an area model to solve 774 ÷ 3.

   b. Draw a number bond to represent this problem.

   c. Record your work using the long division algorithm.

4. a. Draw an area model to solve 1,584 ÷ 2.

   b. Draw a number bond to represent this problem.

   c. Record your work using the long division algorithm.

Lesson 33: Explain the connection of the area model of division to the long division algorithm for three and four digit dividends.

©2015 Great Minds. eureka-math.org
G4-M3-M4-SE-1.3.1-01.2016

Name _____ Date _____

1. Arabelle solved the following division problem by drawing an area model.

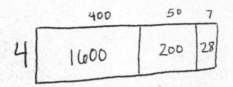

   a. What division problem did she solve?

   b. Show a number bond to represent Arabelle's area model, and represent the total length using the distributive property.

2. a. Solve 816 ÷ 4 using the area model. There is no remainder in this problem.

   b. Draw a number bond and use a written method to record your work from Part (a).

Lesson 33: Explain the connection of the area model of division to the long division algorithm for three and four digit dividends.

©2015 Great Minds. eureka-math.org
G4-M3-M4-SE-1.3.1-01.2016

177

3.   a.   Draw an area model to solve 549 ÷ 3.

    b.   Draw a number bond to represent this problem.

    c.   Record your work using the long division algorithm.

4.   a.   Draw an area model to solve 2,762 ÷ 2.

    b.   Draw a number bond to represent this problem.

    c.   Record your work using the long division algorithm.

Lesson 33:   Explain the connection of the area model of division to the long division algorithm for three and four digit dividends.

©2015 Great Minds. eureka-math.org
G4-M3-M4-SE-1.3.1-01.2016

Name _____ Date _____

1. Use the associative property to rewrite each expression. Solve using disks, and then complete the number sentences.

a. 30 × 24

= (_____ × 10) × 24

= _____ × (10 × 24)

= _____

hundreds	tens	ones

b. 40 × 43

= (4 × 10) × _____

= 4 × (10 × ___)

= _____

thousands	hundreds	tens	ones

c. 30 × 37

= (3 × _____) × _____

= 3 × (10 × _____)

= _____

thousands	hundreds	tens	ones

EUREKA MATH

Lesson 34: Multiply two-digit multiples of 10 by two-digit numbers using a place value chart.

179

©2015 Great Minds. eureka-math.org
G4-M3-M4-SE-1.3.1-01.2016

2. Use the associative property and place value disks to solve.

   a.   20 × 27

   b.   40 × 31

3. Use the associative property without place value disks to solve.

   a.   40 × 34   b.   50 × 43

4. Use the distributive property to solve the following problems. Distribute the second factor.

   a.   40 × 34   b. 60 × 25

©2015 Great Minds. eureka-math.org
G4-M3-M4-SE-1.3.1-01.2016

Name _____ Date _____

1. Use the associative property to rewrite each expression. Solve using disks, and then complete the number sentences.

   a. 20×34

   $= (\text{\_\_\_\_} \times 10) \times 34$

   $= \text{\_\_\_\_} \times (10 \times 34)$

   $= \text{_____}$

hundreds	tens	ones

   b. 30×34

   $= (3 \times 10) \times \text{\_\_\_\_\_}$

   $= 3 \times (10 \times \text{\_\_\_})$

   $= \text{_____}$

thousands	hundreds	tens	ones

   c. 30×42

   $= (3 \times \text{\_\_\_\_}) \times \text{\_\_\_\_\_}$

   $= 3 \times (10 \times \text{\_\_\_\_\_})$

   $= \text{_____}$

thousands	hundreds	tens	ones

Lesson 34: Multiply two-digit multiples of 10 by two-digit numbers using a place value chart.

181

©2015 Great Minds. eureka-math.org
G4-M3-M4-SE-1.3.1-01.2016

2. Use the associative property and place value disks to solve.

    a. 20 × 16   b. 40 × 32

3. Use the associative property without place value disks to solve.

    a. 30 × 21   b. 60 × 42

4. Use the distributive property to solve the following. Distribute the second factor.

    a. 40 × 43   b. 70 × 23

Lesson 34:   Multiply two-digit multiples of 10 by two-digit numbers using a place
                                  value chart.

©2015 Great Minds. eureka-math.org
G4-M3-M4-SE-1.3.1-01.2016

Name _____   Date _____

Use an area model to represent the following expressions. Then, record the partial products and solve.

1. 20 × 22

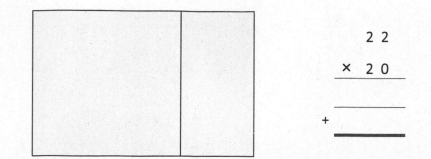

```
      2 2
   ×  2 0
   _____

 + _____
   ━━━━━━━
```

2. 50 × 41

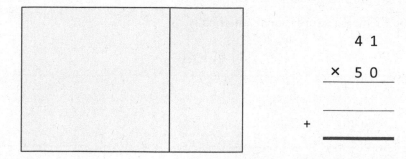

```
      4 1
   ×  5 0
   _____

 + _____
   ━━━━━━━
```

3. 60 × 73

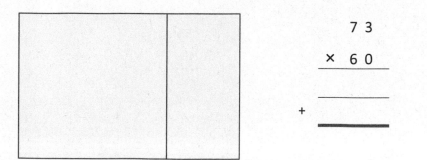

```
      7 3
   ×  6 0
   _____

 + _____
   ━━━━━━━
```

Lesson 35:   Multiply two-digit multiples of 10 by two-digit numbers using the area model.

183

©2015 Great Minds. eureka-math.org
G4-M3-M4-SE-1.3.1-01.2016

Draw an area model to represent the following expressions. Then, record the partial products vertically and solve.

4. 80 × 32

5. 70 × 54

Visualize the area model, and solve the following expressions numerically.

6. 30 × 68

7. 60 × 34

8. 40 × 55

9. 80 × 55

Lesson 35:   Multiply two-digit multiples of 10 by two-digit numbers using the area model.

©2015 Great Minds. eureka-math.org
G4-M3-M4-SE-1.3.1-01.2016

Name _____   Date _____

Use an area model to represent the following expressions. Then, record the partial products and solve.

1. 30 × 17

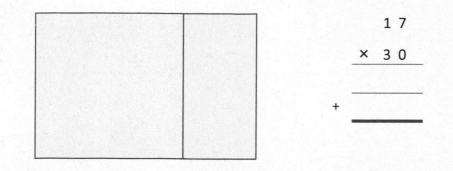

```
      1 7
  ×   3 0
  _____

+ _____
  ═══════
```

2. 40 × 58

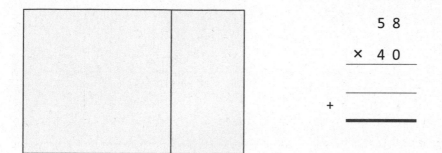

```
      5 8
  ×   4 0
  _____

+ _____
  ═══════
```

3. 50 × 38

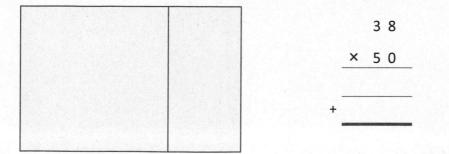

```
      3 8
  ×   5 0
  _____

+ _____
  ═══════
```

Lesson 35:   Multiply two-digit multiples of 10 by two-digit numbers using the area model.

185

©2015 Great Minds. eureka-math.org
G4-M3-M4-SE-1.3.1-01.2016

Draw an area model to represent the following expressions. Then, record the partial products vertically and solve.

4. 60 × 19

5. 20 × 44

Visualize the area model, and solve the following expressions numerically.

6. 20 × 88

7. 30 × 88

8. 70 × 47

9. 80 × 65

Lesson 35: Multiply two-digit multiples of 10 by two-digit numbers using the area model.

©2015 Great Minds. eureka-math.org
G4-M3-M4-SE-1.3.1-01.2016

Name _____   Date _____

1. a. In each of the two models pictured below, write the expressions that determine the area of each of the four smaller rectangles.

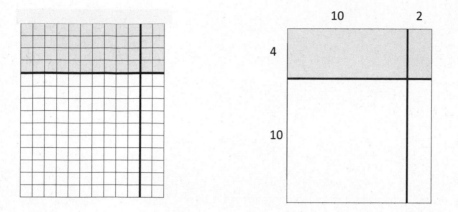

b. Using the distributive property, rewrite the area of the large rectangle as the sum of the areas of the four smaller rectangles. Express first in number form, and then read in unit form.

14 × 12 = (4 × _____) + (4 × _____) + (10 × _____) + (10 × _____)

2. Use an area model to represent the following expression. Record the partial products and solve.

14 × 22

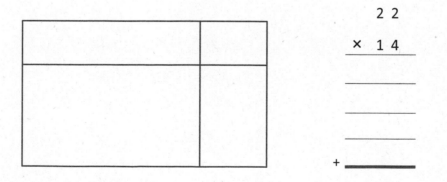

$$\begin{array}{r} 2\ 2 \\ \times\ 1\ 4 \\ \hline \\ \hline \\ \hline \\ +\ \underline{\hspace{2cm}} \end{array}$$

Draw an area model to represent the following expressions. Record the partial products vertically and solve.

3. 25 × 32

4. 35 × 42

Visualize the area model and solve the following numerically using four partial products. (You may sketch an area model if it helps.)

5. 42 × 11

6. 46 × 11

Lesson 36:   Multiply two-digit by two-digit numbers using four partial products.

©2015 Great Minds. eureka-math.org
G4-M3-M4-SE-1.3.1-01.2016

EUREKA
MATH

Name _____   Date _____

1. a. In each of the two models pictured below, write the expressions that determine the area of each of the four smaller rectangles.

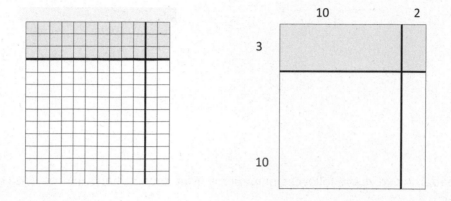

   b. Using the distributive property, rewrite the area of the large rectangle as the sum of the areas of the four smaller rectangles. Express first in number form, and then read in unit form.

   13 × 12 = (3 × _____) + (3 × _____) + (10 × _____) + (10 × _____)

Use an area model to represent the following expression. Record the partial products and solve.

2. 17 × 34

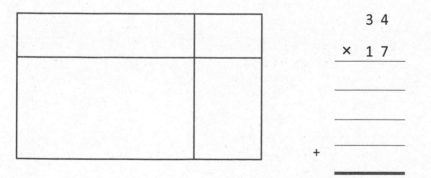

$$\begin{array}{r} 3\ 4 \\ \times\ 1\ 7 \\ \hline \\ \hline \\ +\ \underline{} \\ \hline \end{array}$$

Draw an area model to represent the following expressions. Record the partial products vertically and solve.

3. 45 × 18

4. 45 × 19

Visualize the area model and solve the following numerically using four partial products. (You may sketch an area model if it helps.)

5. 12 × 47

6. 23 × 93

7. 23 × 11

8. 23 × 22

Lesson 36:   Multiply two-digit by two-digit numbers using four partial products.

©2015 Great Minds. eureka-math.org
G4-M3-M4-SE-1.3.1-01.2016

Name _____   Date _____

1. Solve 14 × 12 using 4 partial products and 2 partial products. Remember to think in terms of units as you solve. Write an expression to find the area of each smaller rectangle in the area model.

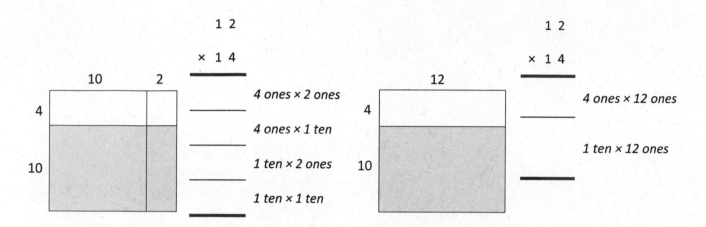

2. Solve 32 × 43 using 4 partial products and 2 partial products. Match each partial product to its area on the models. Remember to think in terms of units as you solve.

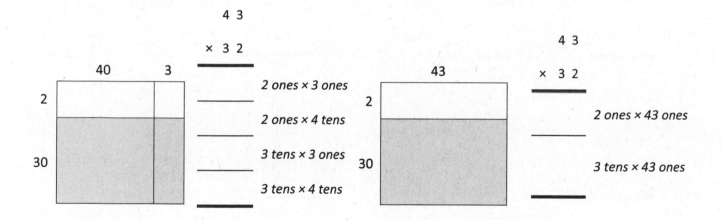

Lesson 37:   Transition from four partial products to the standard algorithm for
two-digit by two digit multiplication.

191

©2015 Great Minds. eureka-math.org
G4-M3-M4-SE-1.3.1-01.2016

3. Solve 57 × 15 using 2 partial products. Match each partial product to its rectangle on the area model.

4. Solve the following using 2 partial products. Visualize the area model to help you.

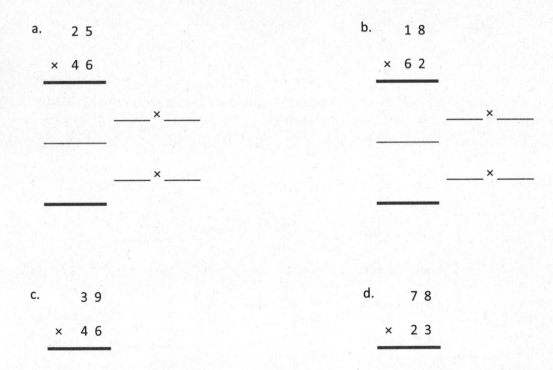

a.
    2 5
  × 4 6
  _____

    _____ × _____

  _____

    _____ × _____

  _____

b.
    1 8
  × 6 2
  _____

    _____ × _____

  _____

    _____ × _____

  _____

c.
    3 9
  × 4 6
  _____

d.
    7 8
  × 2 3
  _____

Lesson 37:   Transition from four partial products to the standard algorithm for
                two-digit by two digit multiplication.

©2015 Great Minds. eureka-math.org
G4-M3-M4-SE-1.3.1-01.2016

EUREKA
MATH

Name _____   Date _____

1. Solve 26 × 34 using 4 partial products and 2 partial products. Remember to think in terms of units as you solve. Write an expression to find the area of each smaller rectangle in the area model.

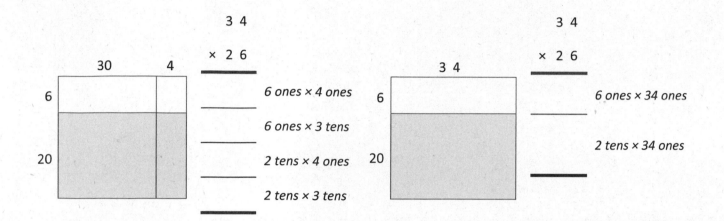

2. Solve using 4 partial products and 2 partial products. Remember to think in terms of units as you solve. Write an expression to find the area of each smaller rectangle in the area model.

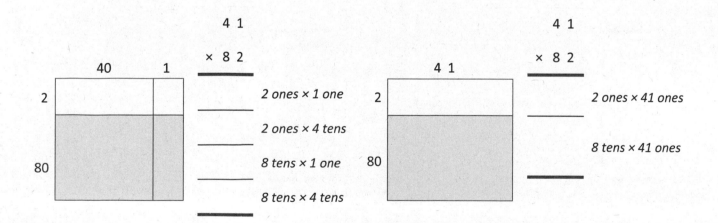

Lesson 37:   Transition from four partial products to the standard algorithm for two-digit by two digit multiplication.   193

©2015 Great Minds. eureka-math.org
G4-M3-M4-SE-1.3.1-01.2016

3. Solve 52 × 26 using 2 partial products and an area model. Match each partial product to its area on the model.

4. Solve the following using 2 partial products. Visualize the area model to help you.

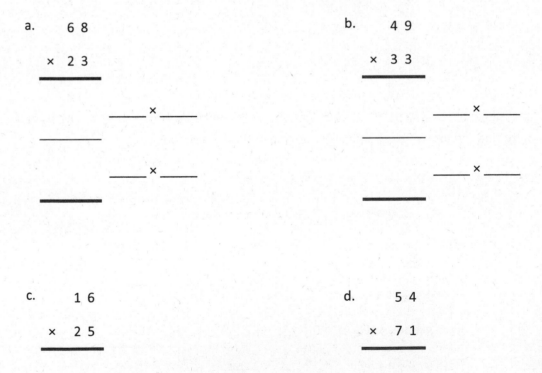

a.   6 8
    × 2 3
    ─────────

        ____ × ____

    _____

        ____ × ____
    ─────────

b.   4 9
    × 3 3
    ─────────

        ____ × ____

    _____

        ____ × ____
    ─────────

c.   1 6
    × 2 5
    ─────────

d.   5 4
    × 7 1
    ─────────

Lesson 37:   Transition from four partial products to the standard algorithm for two-digit by two digit multiplication.

©2015 Great Minds. eureka-math.org
G4-M3-M4-SE-1.3.1-01.2016

EUREKA
MATH™

Name _____ Date _____

1. Express 23 × 54 as two partial products using the distributive property. Solve.

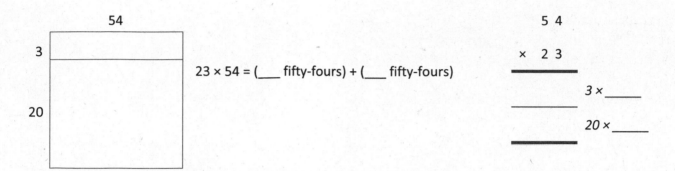

23 × 54 = (____ fifty-fours) + (____ fifty-fours)

```
      5 4
  ×   2 3
  _____
          3 × _____
  _____
         20 × _____
  _____
```

2. Express 46 × 54 as two partial products using the distributive property. Solve.

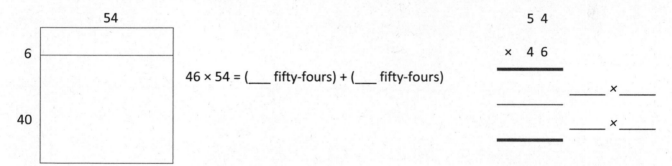

46 × 54 = (____ fifty-fours) + (____ fifty-fours)

```
      5 4
  ×   4 6
  _____
          _____ × _____
  _____
          _____ × _____
  _____
```

3. Express 55 × 47 as two partial products using the distributive property. Solve.

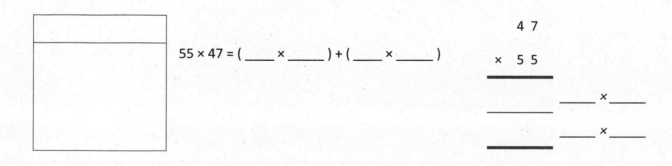

55 × 47 = (____ × _____) + (____ × _____)

```
      4 7
  ×   5 5
  _____
          _____ × _____
  _____
          _____ × _____
  _____
```

Lesson 38: Transition from four partial products to the standard algorithm for two-digit by two digit multiplication.

195

©2015 Great Minds. eureka-math.org
G4-M3-M4-SE-1.3.1-01.2016

4. Solve the following using 2 partial products.

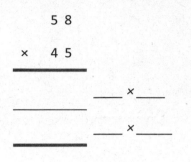

5. Solve using the multiplication algorithm.

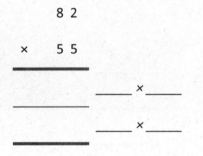

6. 53 × 63

7. 84 × 73

Lesson 38:   Transition from four partial products to the standard algorithm for two-digit by two digit multiplication.

©2015 Great Minds. eureka-math.org
G4-M3-M4-SE-1.3.1-01.2016

Name _____ Date _____

1. Express 26 × 43 as two partial products using the distributive property. Solve.

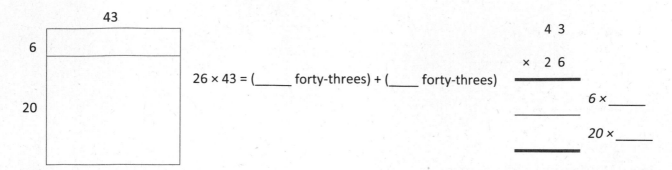

43

6

20

26 × 43 = (_____ forty-threes) + (_____ forty-threes)

```
      4 3
    × 2 6
    ───────
              6 × _____
    ───────
             20 × _____
    ───────
```

2. Express 47 × 63 as two partial products using the distributive property. Solve.

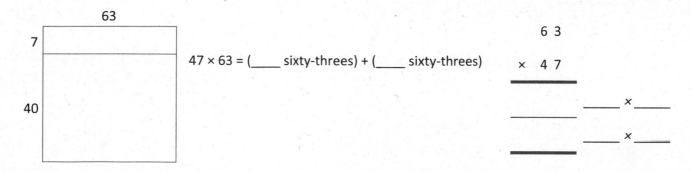

63

7

40

47 × 63 = (_____ sixty-threes) + (_____ sixty-threes)

```
      6 3
    × 4 7
    ───────
              _____ × _____
    ───────
              _____ × _____
    ───────
```

3. Express 54 × 67 as two partial products using the distributive property. Solve.

54 × 67 = (___ × _____) + (___ × _____)

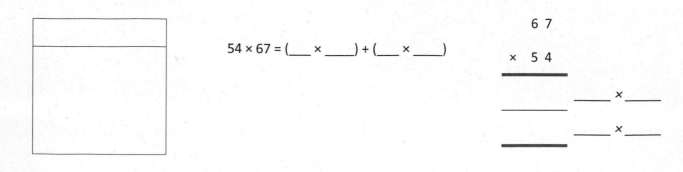

```
      6 7
    × 5 4
    ───────
              _____ × _____
    ───────
              _____ × _____
    ───────
```

  Lesson 38:   Transition from four partial products to the standard algorithm for   197
             two-digit by two digit multiplication.

©2015 Great Minds. eureka-math.org
G4-M3-M4-SE-1.3.1-01.2016

4. Solve the following using two partial products.

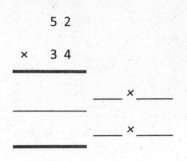

```
        5 2
    ×   3 4
  ─────────      ____ × ____
  ─────────
                 ____ × ____
  ═════════
```

5. Solve using the multiplication algorithm.

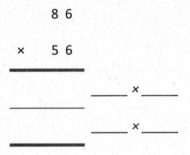

```
        8 6
    ×   5 6
  ─────────      _____ × _____
  ─────────
                 _____ × _____
  ═════════
```

6. 54 × 52   7. 44 × 76

Lesson 38:   Transition from four partial products to the standard algorithm for two-digit by two digit multiplication.

©2015 Great Minds. eureka-math.org
G4-M3-M4-SE-1.3.1-01.2016

EUREKA MATH

8. 63 × 63

9. 68 × 79

Lesson 38:   Transition from four partial products to the standard algorithm for
                 two-digit by two digit multiplication.

199

©2015 Great Minds. eureka-math.org
G4-M3-M4-SE-1.3.1-01.2016

This page intentionally left blank

Student Edition

Eureka Math

Grade 4
Module 4

Special thanks go to the Gordon A. Cain Center and to the Department of Mathematics at Louisiana State University for their support in the development of *Eureka Math*.

For a free *Eureka Math* Teacher Resource Pack, Parent Tip Sheets, and more please visit www.Eureka.tools

Published by the non-profit Great Minds

Copyright © 2015 Great Minds. No part of this work may be reproduced, sold, or commercialized, in whole or in part, without written permission from Great Minds. Non-commercial use is licensed pursuant to a Creative Commons Attribution-NonCommercial-ShareAlike 4.0 license; for more information, go to http://greatminds.net/maps/math/copyright. "Great Minds" and "Eureka Math" are registered trademarks of Great Minds.

Printed in the U.S.A.
This book may be purchased from the publisher at eureka-math.org
10 9

ISBN 978-1-63255-304-1

Name _____   Date _____

1. Use the following directions to draw a figure in the box to the right.

    a. Draw two points: A and B.

    b. Use a straightedge to draw $\overrightarrow{AB}$.

    c. Draw a new point that is not on $\overrightarrow{AB}$. Label it C.

    d. Draw $\overline{AC}$.

    e. Draw a point not on $\overrightarrow{AB}$ or $\overline{AC}$. Call it D.

    f. Construct $\overleftrightarrow{CD}$.

    g. Use the points you've already labeled to name one

        angle. _____

2. Use the following directions to draw a figure in the box to the right.

    a. Draw two points: A and B.

    b. Use a straightedge to draw $\overline{AB}$.

    c. Draw a new point that is not on $\overline{AB}$. Label it C.

    d. Draw $\overline{BC}$.

    e. Draw a new point that is not on $\overline{AB}$ or $\overline{BC}$.

        Label it D.

    f. Construct $\overleftrightarrow{AD}$.

    g. Identify $\angle DAB$ by drawing an arc to indicate the

        position of the angle.

    h. Identify another angle by referencing points that

        you have already drawn. _____

Lesson 1:   Identify and draw points, lines, line segments, rays, and angles.
             Recognize them in various contexts and familiar figures.

©2015 Great Minds. eureka-math.org
G4-M3-M4-SE-1.3.1-01.2016

1

3. a. Observe the familiar figures below. Label some points on each figure.

    b. Use those points to label and name representations of each of the following in the table below: ray, line, line segment, and angle. Extend segments to show lines and rays.

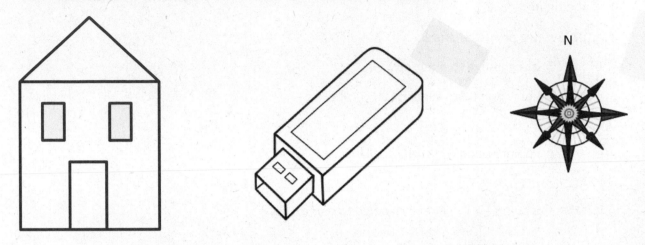

	House	Flash drive	Compass rose
Ray			
Line			
Line segment			
Angle			

Extension: Draw a familiar figure. Label it with points, and then identify rays, lines, line segments, and angles as applicable.

Lesson 1:   Identify and draw points, lines, line segments, rays, and angles. Recognize them in various contexts and familiar figures.

©2015 Great Minds. eureka-math.org
G4-M3-M4-SE-1.3.1-01.2016

Name _____   Date _____

1. Use the following directions to draw a figure in the box to the right.

    a. Draw two points: W and X.

    b. Use a straightedge to draw $\overleftrightarrow{WX}$.

    c. Draw a new point that is not on $\overleftrightarrow{WX}$. Label it Y.

    d. Draw $\overline{WY}$.

    e. Draw a point not on $\overleftrightarrow{WX}$ or $\overline{WY}$. Call it Z.

    f. Construct $\overleftrightarrow{YZ}$.

    g. Use the points you've already labeled to name one angle. _____

2. Use the following directions to draw a figure in the box to the right.

    a. Draw two points: W and X.

    b. Use a straightedge to draw $\overline{WX}$.

    c. Draw a new point that is not on $\overline{WX}$. Label it Y.

    d. Draw $\overrightarrow{WY}$.

    e. Draw a new point that is not on $\overrightarrow{WY}$ or on the line containing $\overline{WX}$. Label it Z.

    f. Construct $\overleftrightarrow{WZ}$.

    g. Identify $\angle ZWX$ by drawing an arc to indicate the position of the angle.

    h. Identify another angle by referencing points that you have already drawn. _____

Lesson 1:   Identify and draw points, lines, line segments, rays, and angles.
Recognize them in various contexts and familiar figures.

3

©2015 Great Minds. eureka-math.org
G4-M3-M4-SE-1.3.1-01.2016

3. a. Observe the familiar figures below. Label some points on each figure.

    b. Use those points to label and name representations of each of the following in the table below: ray, line, line segment, and angle. Extend segments to show lines and rays.

	Clock	Die	Number line
Ray			
Line			
Line segment			
Angle			

Extension: Draw a familiar figure. Label it with points, and then identify rays, lines, line segments, and angles as applicable.

Lesson 1:   Identify and draw points, lines, line segments, rays, and angles.
                                Recognize them in various contexts and familiar figures.

©2015 Great Minds. eureka-math.org
G4-M3-M4-SE-1.3.1-01.2016

Name _____   Date _____

1. Use the right angle template that you made in class to determine if each of the following angles is greater than, less than, or equal to a right angle. Label each as *greater than*, *less than*, or *equal to*, and then connect each angle to the correct label of acute, right, or obtuse.
    The first one has been completed for you.

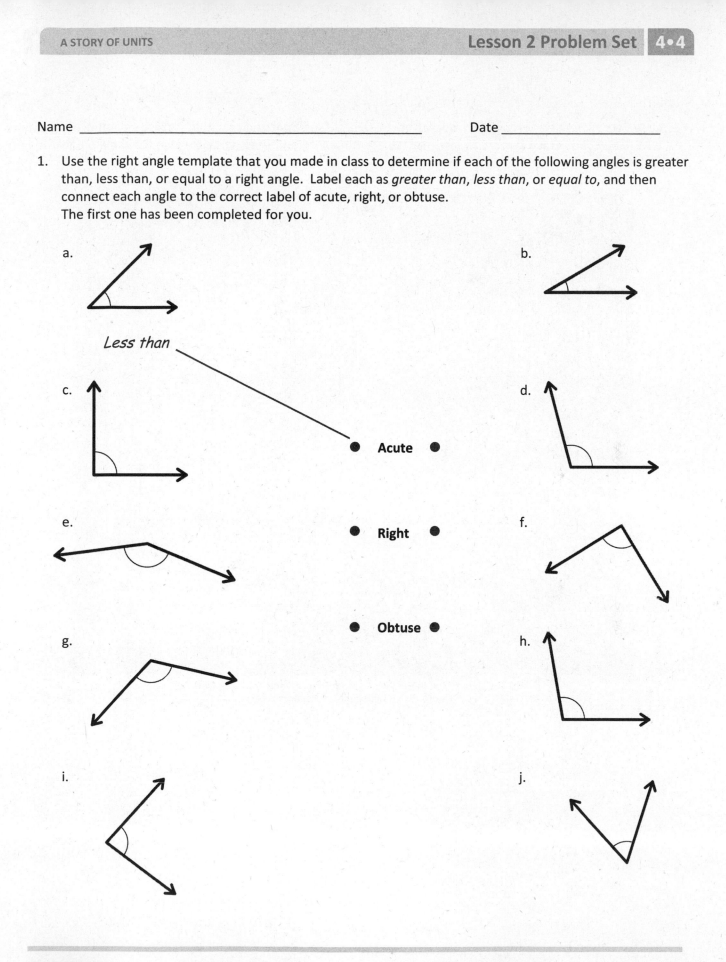

a.

Less than

b.

c.

• **Acute** •

d.

e.

• **Right** •

f.

• **Obtuse** •

g.

h.

i.

j.

EUREKA MATH

Lesson 2:   Use right angles to determine whether angles are equal to, greater than, or less than right angles. Draw right, obtuse, and acute angles.

5

©2015 Great Minds. eureka-math.org
G4-M3-M4-SE-1.3.1-01.2016

2. Use your right angle template to identify acute, obtuse, and right angles within Picasso's painting *Factory, Horta de Ebbo*. Trace at least two of each, label with points, and then name them in the table below the painting.

© 2013 Estate of Pablo Picasso / Artists Rights Society (ARS), New York
Photo: Erich Lessing / Art Resource, NY.

Acute angle		
Obtuse angle		
Right angle		

Lesson 2:   Use right angles to determine whether angles are equal to, greater than, or less than right angles. Draw right, obtuse, and acute angles.

©2015 Great Minds. eureka-math.org
G4-M3-M4-SE-1.3.1-01.2016

3. Construct each of the following using a straightedge and the right angle template that you created. Explain the characteristics of each by comparing the angle to a right angle. Use the words *greater than*, *less than,* or *equal to* in your explanations.

   a. Acute angle

   b. Right angle

   c. Obtuse angle

Lesson 2:   Use right angles to determine whether angles are equal to, greater than, or less than right angles. Draw right, obtuse, and acute angles.

©2015 Great Minds. eureka-math.org
G4-M3-M4-SE-1.3.1-01.2016

7

Name _____ Date _____

1. Use the right angle template that you made in class to determine if each of the following angles is greater than, less than, or equal to a right angle. Label each as *greater than*, *less than*, or *equal to*, and then connect each angle to the correct label of acute, right, or obtuse. The first one has been completed for you.

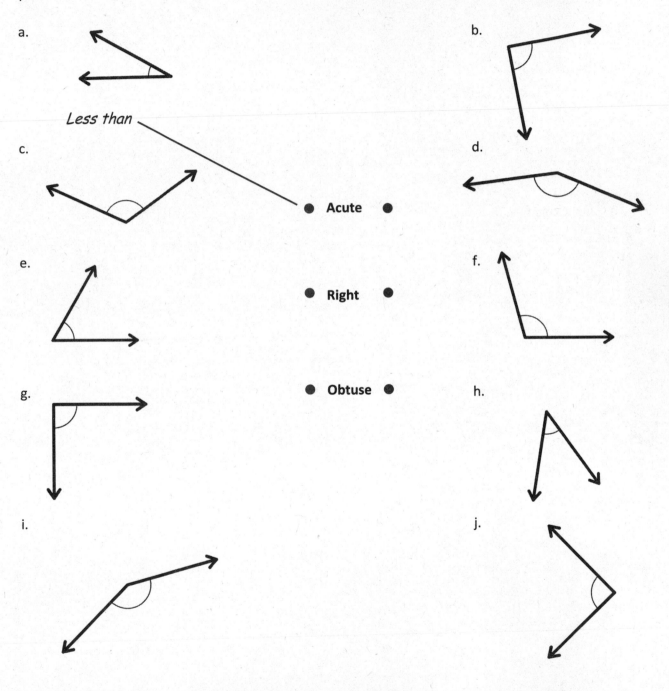

a.

Less than

b.

c.

● **Acute** ●

d.

e.

● **Right** ●

f.

● **Obtuse** ●

g.

h.

i.

j.

Lesson 2: Use right angles to determine whether angles are equal to, greater than, or less than right angles. Draw right, obtuse, and acute angles.

©2015 Great Minds. eureka-math.org
G4-M3-M4-SE-1.3.1-01.2016

2. Use your right angle template to identify acute, obtuse, and right angles within this painting. Trace at least two of each, label with points, and then name them in the table below the painting.

Acute angle		
Obtuse angle		
Right angle		

Lesson 2:   Use right angles to determine whether angles are equal to, greater than, or less than right angles. Draw right, obtuse, and acute angles.

©2015 Great Minds. eureka-math.org
G4-M3-M4-SE-1.3.1-01.2016

9

3. Construct each of the following using a straightedge and the right angle template that you created. Explain the characteristics of each by comparing the angle to a right angle. Use the words *greater than*, *less than,* or *equal to* in your explanations.

a. Acute angle

b. Right angle

c. Obtuse angle

Lesson 2:   Use right angles to determine whether angles are equal to, greater than, or less than right angles. Draw right, obtuse, and acute angles.

©2015 Great Minds. eureka-math.org
G4-M3-M4-SE-1.3.1-01.2016

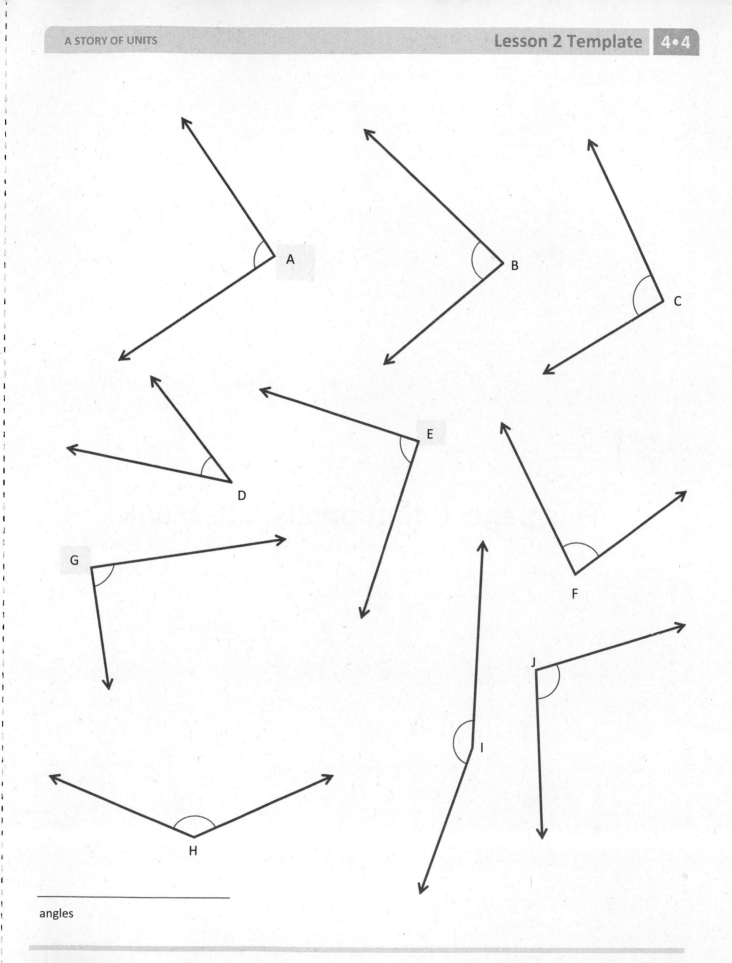

angles

EUREKA MATH™

Lesson 2:   Use right angles to determine whether angles are equal to, greater than, or less than right angles. Draw right, obtuse, and acute angles.

©2015 Great Minds. eureka-math.org
G4-M3-M4-SE-1.3.1-01.2016

11

This page intentionally left blank

Name _____   Date _____

1. On each object, trace at least one pair of lines that appear to be perpendicular.

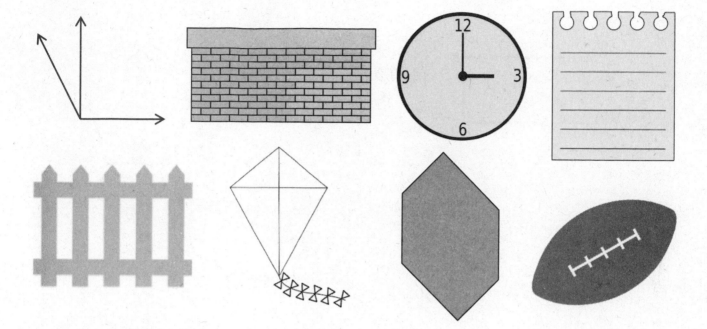

2. How do you know if two lines are perpendicular?

intersect at right angle.

3. In the square and triangular grids below, use the given segments in each grid to draw a segment that is perpendicular using a straightedge.

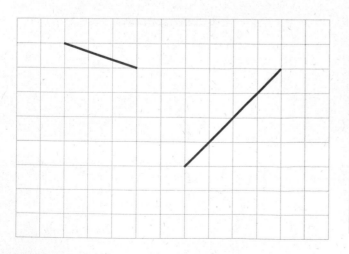
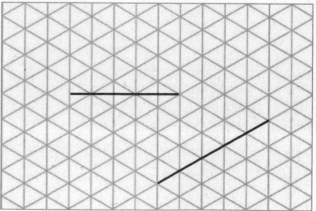

4. Use the right angle template that you created in class to determine which of the following figures have a right angle. Mark each right angle with a small square. For each right angle you find, name the corresponding pair of perpendicular sides. (Problem 4(a) has been started for you.)

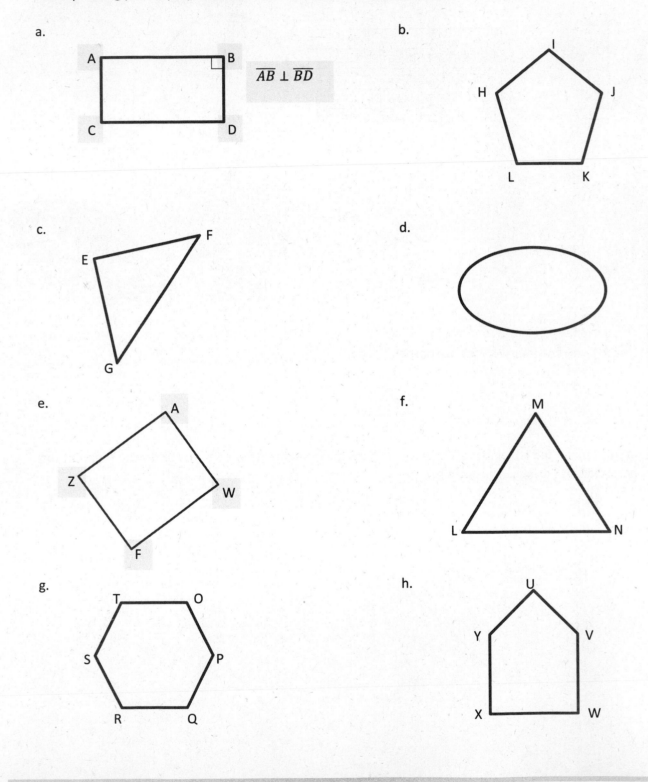

a.

$\overline{AB} \perp \overline{BD}$

b.

c.

d.

e.

f.

g.

h.

Lesson 3: Identify, define, and draw perpendicular lines.

EUREKA MATH

©2015 Great Minds. eureka-math.org
G4-M3-M4-SE-1.3.1-01.2016

5. Mark each right angle on the following figure with a small square. (Note: A right angle does not have to be inside the figure.) How many pairs of perpendicular sides does this figure have?

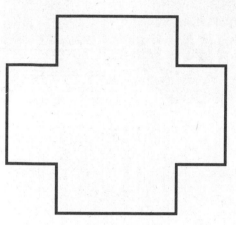

6. True or false? Shapes that have at least one right angle also have at least one pair of perpendicular sides. Explain your thinking.

©2015 Great Minds. eureka-math.org
G4-M3-M4-SE-1.3.1-01.2016

Name _____ Date _____

1. On each object, trace at least one pair of lines that appear to be perpendicular.

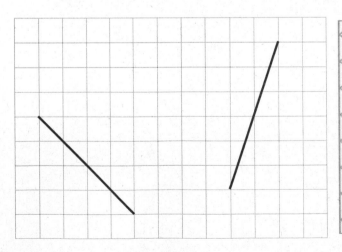

2. How do you know if two lines are perpendicular?

3. In the square and triangular grids below, use the given segments in each grid to draw a segment that is perpendicular. Use a straightedge.

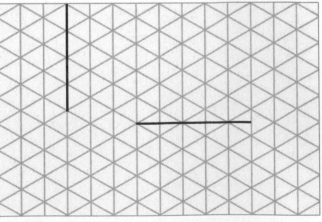

Lesson 3:   Identify, define, and draw perpendicular lines.

©2015 Great Minds. eureka-math.org
G4-M3-M4-SE-1.3.1-01.2016

4. Use the right angle template that you created in class to determine which of the following figures have a right angle. Mark each right angle with a small square. For each right angle you find, name the corresponding pair of perpendicular sides. (Problem 4(a) has been started for you.)

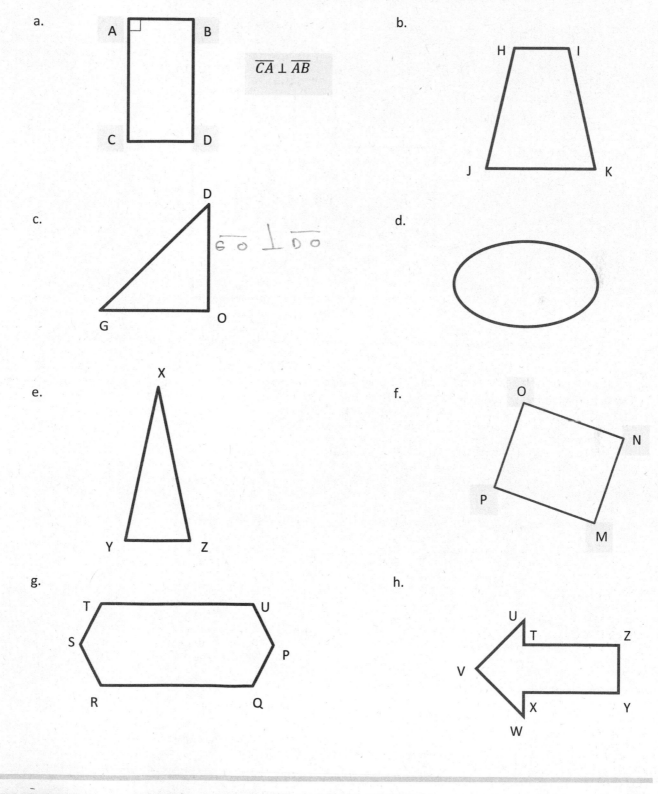

a.

$$\overline{CA} \perp \overline{AB}$$

b.

c.

d.

e.

f.

g.

h.

5. Use your right angle template as a guide, and mark each right angle in the following figure with a small square. (Note: A right angle does not have to be inside the figure.) How many pairs of perpendicular sides does this figure have?

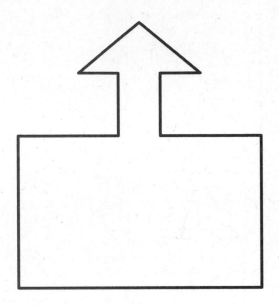

6. True or false? Shapes that have no right angles also have no perpendicular segments. Draw some figures to help explain your thinking.

©2015 Great Minds. eureka-math.org
G4-M3-M4-SE-1.3.1-01.2016

Name _____   Date _____

1. On each object, trace at least one pair of lines that appear to be parallel.

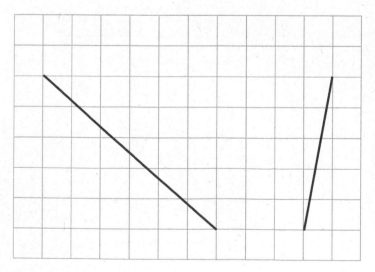

2. How do you know if two lines are parallel?

3. In the square and triangular grids below, use the given segments in each grid to draw a segment that is parallel using a straightedge.

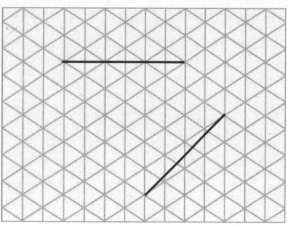

4. Determine which of the following figures have sides that are parallel by using a straightedge and the right angle template that you created. Circle the letter of the shapes that have at least one pair of parallel sides. Mark each pair of parallel sides with arrowheads, and then identify the parallel sides with a statement modeled after the one in 4(a).

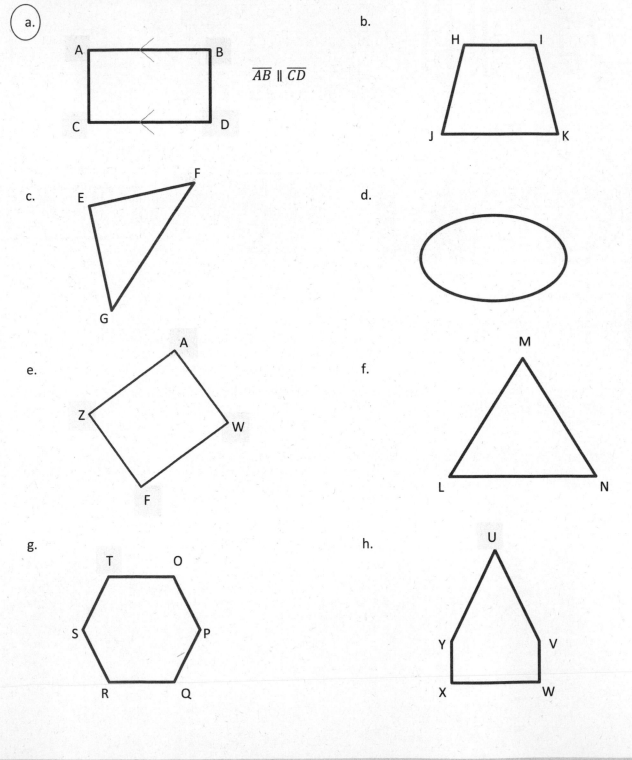

a.

$\overline{AB} \parallel \overline{CD}$

b.

c.

d.

e.

f.

g.

h.

Lesson 4:   Identify, define, and draw parallel lines.

©2015 Great Minds. eureka-math.org
G4-M3-M4-SE-1.3.1-01.2016

5. True or false? A triangle cannot have sides that are parallel. Explain your thinking.

6. Explain why $\overline{AB}$ and $\overline{CD}$ are parallel, but $\overline{EF}$ and $\overline{GH}$ are not.

7. Draw a line using your straightedge. Now, use your right angle template and straightedge to construct a line parallel to the first line you drew.

Name _____   Date _____

1. On each object, trace at least one pair of lines that appear to be parallel.

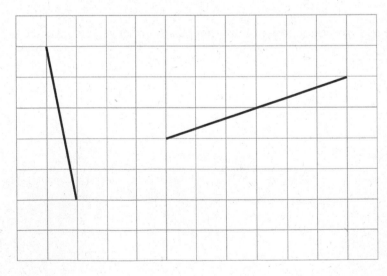

2. How do you know if two lines are parallel?

3. In the square and triangular grids below, use the given segments in each grid to draw a segment that is parallel using a straightedge.

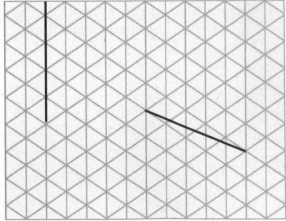

©2015 Great Minds. eureka-math.org
G4-M3-M4-SE-1.3.1-01.2016

4. Determine which of the following figures have sides that are parallel by using a straightedge and the right angle template that you created. Circle the letter of the shapes that have at least one pair of parallel sides. Mark each pair of parallel sides with arrows, and then identify the parallel sides with a statement modeled after the one in 4(a).

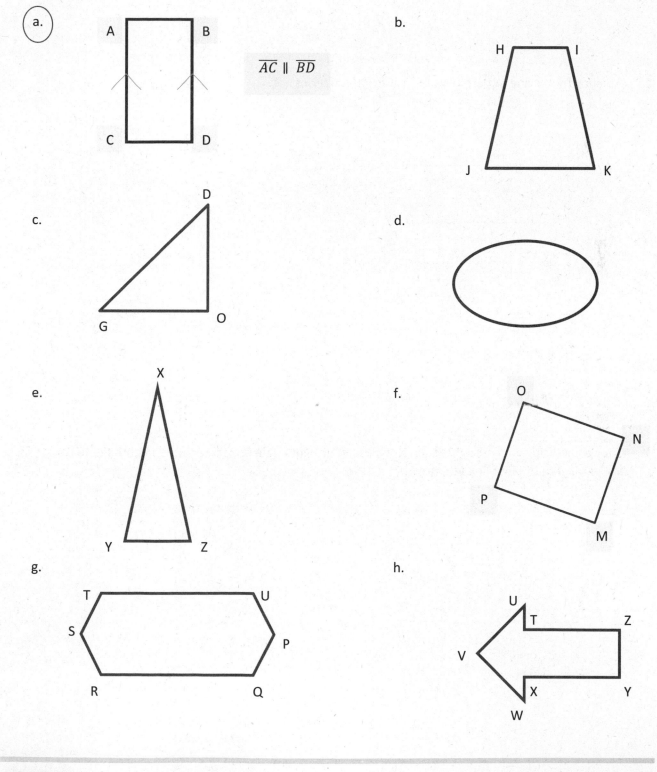

a.

$\overline{AC} \parallel \overline{BD}$

b.

c.

d.

e.

f.

g.

h.

EUREKA MATH™

©2015 Great Minds. eureka-math.org
G4-M3-M4-SE-1.3.1-01.2016

5. True or false? All shapes with a right angle have sides that are parallel. Explain your thinking.

6. Explain why $\overline{AB}$ and $\overline{CD}$ are parallel, but $\overline{EF}$ and $\overline{GH}$ are not.

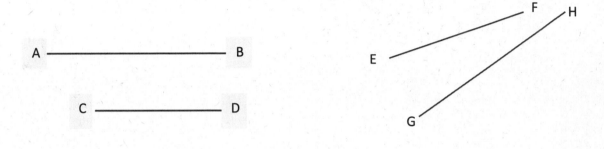

7. Draw a line using your straightedge. Now, use your right angle template and straightedge to construct a line parallel to the first line you drew.

©2015 Great Minds. eureka-math.org
G4-M3-M4-SE-1.3.1-01.2016

Name _____   Date _____

1. Make a list of the measures of the benchmark angles you drew, starting with Set A.
    Round each angle measure to the nearest 5°. Both sets have been started for you.

    a. Set A: 45°, 90°,

    b. Set B: 30°, 60°,

2. Circle any angle measures that appear on both lists. What do you notice about them?

3. List the angle measures from Problem 1 that are acute. Trace each angle with your finger as you say its measurement.

4. List the angle measures from Problem 1 that are obtuse. Trace each angle with your finger as you say its measurement.

Lesson 5:   Use a circular protractor to understand a 1-degree angle as $\frac{1}{360}$ of a turn. Explore benchmark angles using the protractor.

©2015 Great Minds. eureka-math.org
G4-M3-M4-SE-1.3.1-01.2016

25

5. We found out today that 1° is $\frac{1}{360}$ of a whole turn. It is 1 out of 360°. That means a 2° angle is $\frac{2}{360}$ of a whole turn. What fraction of a whole turn is each of the benchmark angles you listed in Problem 1?

6. How many 45° angles does it take to make a full turn?

7. How many 30° angles does it take to make a full turn?

8. If you didn't have a protractor, how could you reconstruct a quarter of it from 0° to 90°?

Lesson 5:   Use a circular protractor to understand a 1-degree angle as $\frac{1}{360}$ of a turn. Explore benchmark angles using the protractor.

©2015 Great Minds. eureka-math.org
G4-M3-M4-SE-1.3.1-01.2016

EUREKA
MATH™

Name _____   Date _____

1. Identify the measures of the following angles.

a.

b.

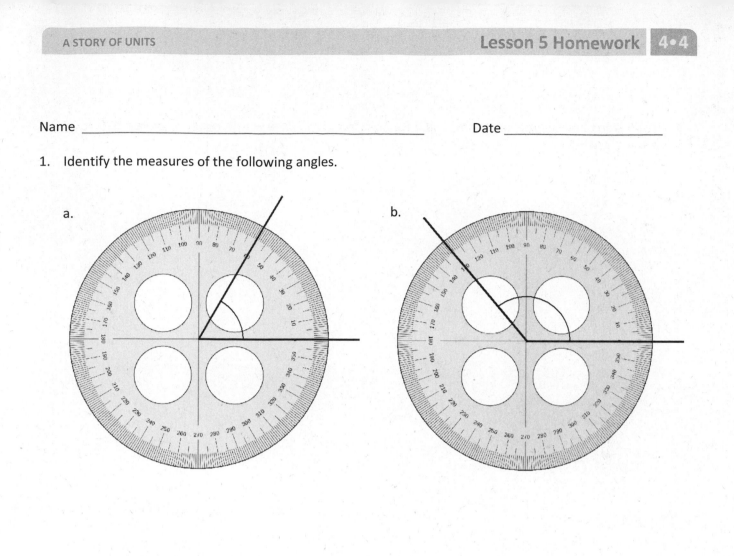

c.

d.

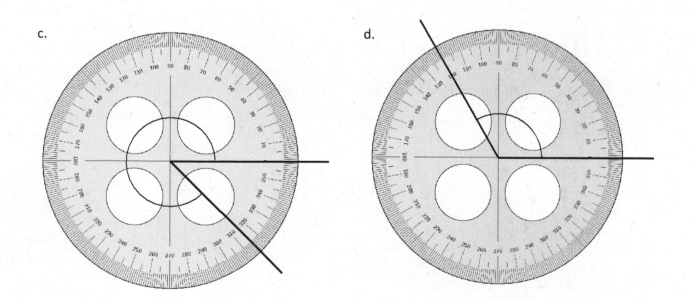

EUREKA
MATH™

Lesson 5:   Use a circular protractor to understand a 1-degree angle as $\frac{1}{360}$ of a
turn. Explore benchmark angles using the protractor.

©2015 Great Minds. eureka-math.org
G4-M3-M4-SE-1.3.1-01.2016

27

2. If you didn't have a protractor, how could you construct one? Use words, pictures, or numbers to explain in the space below.

Lesson 5:   Use a circular protractor to understand a 1-degree angle as $\frac{1}{360}$ of a turn. Explore benchmark angles using the protractor.

©2015 Great Minds. eureka-math.org
G4-M3-M4-SE-1.3.1-01.2016

Name _____   Date _____

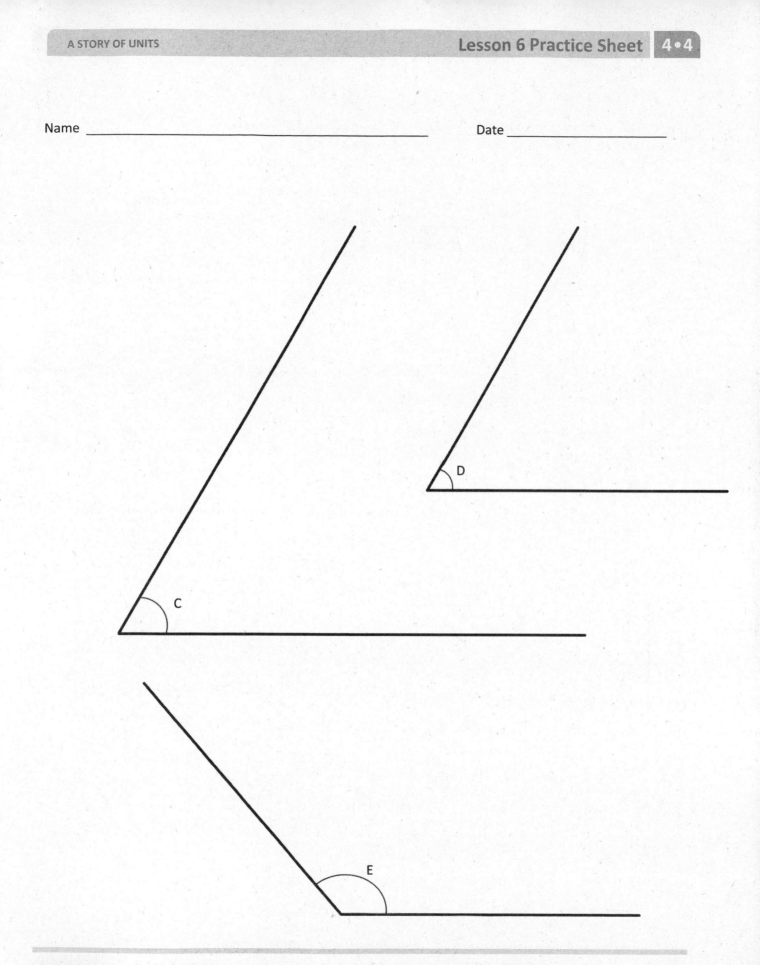

EUREKA
MATH™

Lesson 6:   Use varied protractors to distinguish angle measure from length
             measurement.

©2015 Great Minds. eureka-math.org
G4-M3-M4-SE-1.3.1-01.2016

29

Name _____   Date _____

1. Use a protractor to measure the angles, and then record the measurements in degrees.

a.

b.

c.

d.

Lesson 6:   Use varied protractors to distinguish angle measure from length
measurement.

©2015 Great Minds. eureka-math.org
G4-M3-M4-SE-1.3.1-01.2016

e.

f.

g.

h.

i.

j.

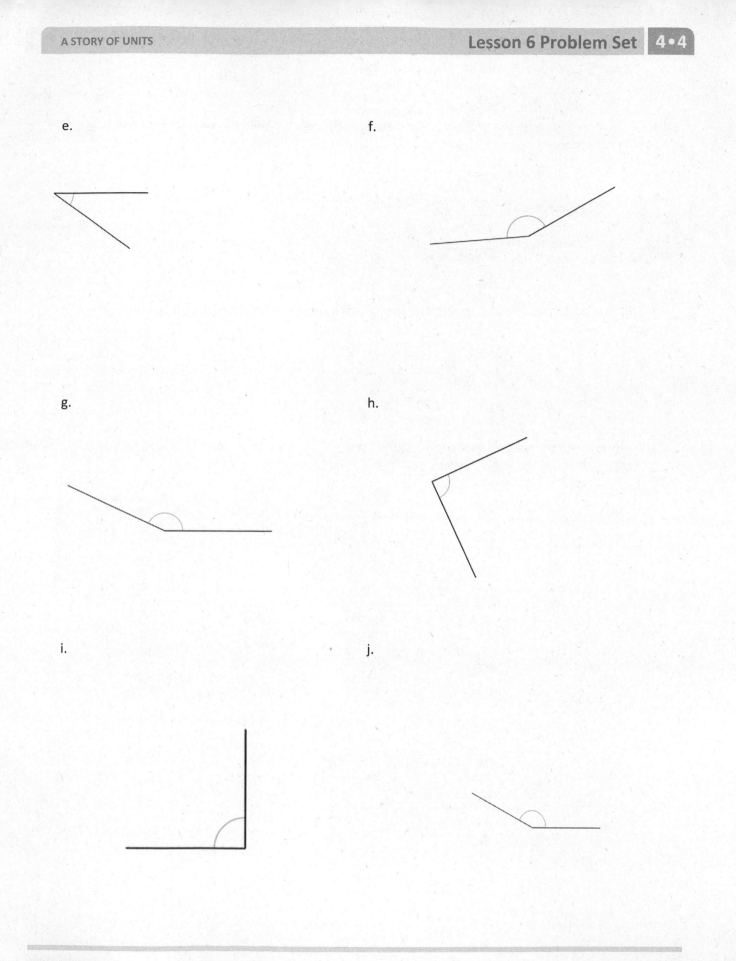

EUREKA MATH™

Lesson 6:   Use varied protractors to distinguish angle measure from length
measurement.

31

©2015 Great Minds. eureka-math.org
G4-M3-M4-SE-1.3.1-01.2016

2. a. Use three different-size protractors to measure the angle. Extend the lines as needed using a straightedge.

Protractor #1: _____ °

Protractor #2: _____ °

Protractor #3: _____ °

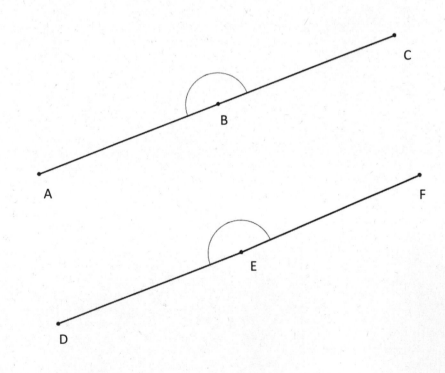

    b. What do you notice about the measurement of the above angle using each of the protractors?

3. Use a protractor to measure each angle. Extend the length of the segments as needed. When you extend the segments, does the angle measure stay the same? Explain how you know.

a.

b.

Lesson 6:   Use varied protractors to distinguish angle measure from length measurement.

©2015 Great Minds. eureka-math.org
G4-M3-M4-SE-1.3.1-01.2016

Name _____ Date _____

1. Use a protractor to measure the angles, and then record the measurements in degrees.

a.

b.

c.

d.

Lesson 6:   Use varied protractors to distinguish angle measure from length
              measurement.

©2015 Great Minds. eureka-math.org
G4-M3-M4-SE-1.3.1-01.2016

33

e.

f.

g.

h.

i.

j.

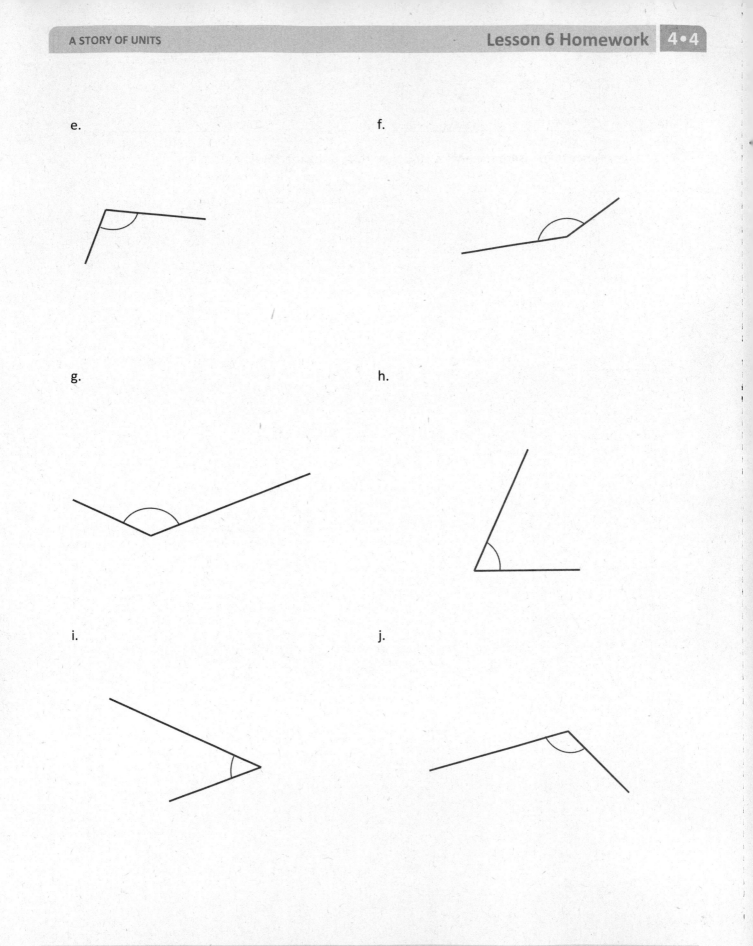

Lesson 6: Use varied protractors to distinguish angle measure from length measurement.

©2015 Great Minds. eureka-math.org
G4-M3-M4-SE-1.3.1-01.2016

2. Using the green and red circle cutouts from today's lesson, explain to someone at home how the cutouts can be used to show that the angle measures are the same even though the circles are different sizes. Write words to explain what you told him or her.

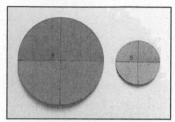

3. Use a protractor to measure each angle. Extend the length of the segments as needed. When you extend the segments, does the angle measure stay the same? Explain how you know.

a.

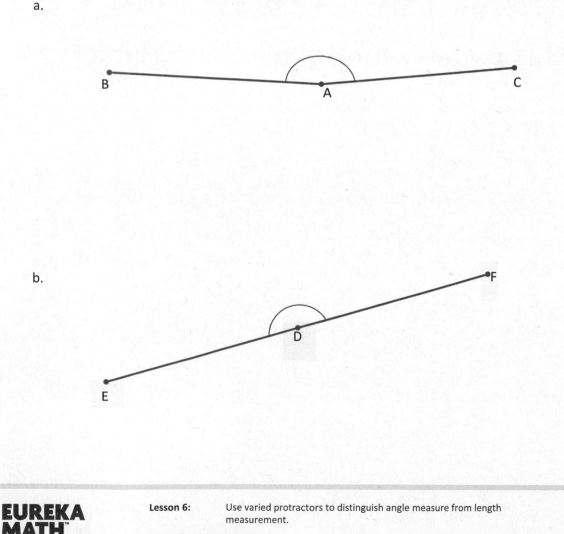

b.

Lesson 6:   Use varied protractors to distinguish angle measure from length measurement.

35

EUREKA
MATH™

©2015 Great Minds. eureka-math.org
G4-M3-M4-SE-1.3.1-01.2016

This page intentionally left blank

Name _____   Date _____

Figure 1

Figure 2

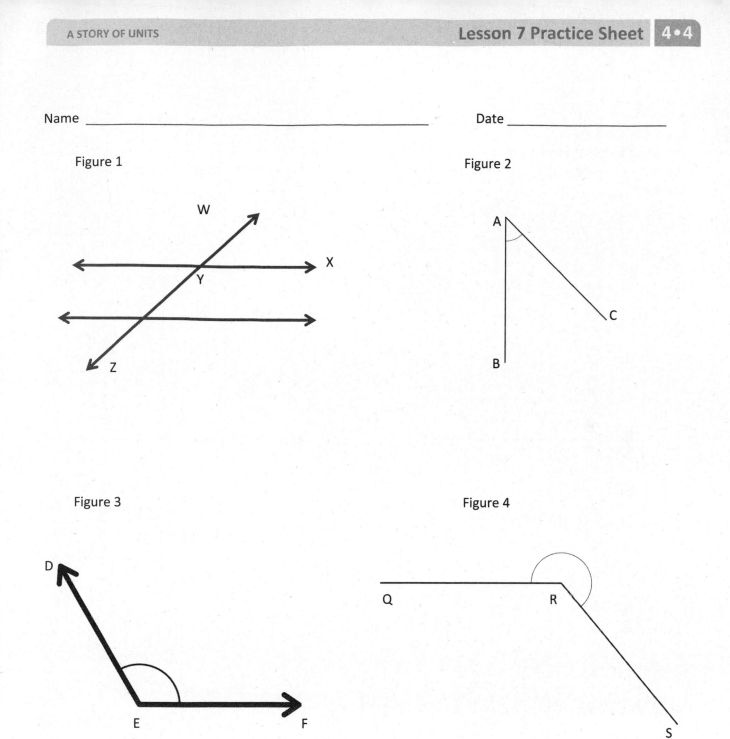

Figure 3

Figure 4

Lesson 7: Measure and draw angles. Sketch given angle measures, and verify with a protractor.

©2015 Great Minds. eureka-math.org
G4-M3-M4-SE-1.3.1-01.2016

37

Name _____ Date _____

Construct angles that measure the given number of degrees. For Problems 1–4, use the ray shown as one of the rays of the angle with its endpoint as the vertex of the angle. Draw an arc to indicate the angle that was measured.

1. 30°

2. 65°

3. 115°

4. 135°

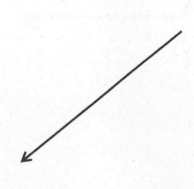

Lesson 7: Measure and draw angles. Sketch given angle measures, and verify with a protractor.

©2015 Great Minds. eureka-math.org
G4-M3-M4-SE-1.3.1-01.2016

5.   5°   6.   175°

7.   27°   8.   117°

9.   48°   10. 132°

Lesson 7:   Measure and draw angles. Sketch given angle measures, and verify
              with a protractor.

©2015 Great Minds. eureka-math.org
G4-M3-M4-SE-1.3.1-01.2016

39

Name _____　　Date _____

Construct angles that measure the given number of degrees. For Problems 1–4, use the ray shown as one of the rays of the angle with its endpoint as the vertex of the angle. Draw an arc to indicate the angle that was measured.

1. 25°　　　　　　　　　　　　　　　　　　　　2. 85°

3. 140°　　　　　　　　　　　　　　　　　　　4. 83°

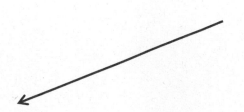

　　　　Lesson 7:　　Measure and draw angles. Sketch given angle measures, and verify with a protractor.

©2015 Great Minds. eureka-math.org
G4-M3-M4-SE-1.3.1-01.2016

EUREKA
MATH™

5. 108°   6. 72°

7. 25°   8. 155°

9. 45°   10. 135°

Lesson 7:   Measure and draw angles. Sketch given angle measures, and verify   41
              with a protractor.

©2015 Great Minds. eureka-math.org
G4-M3-M4-SE-1.3.1-01.2016

This page intentionally left blank

Name _____ Date _____

1. Joe, Steve, and Bob stood in the middle of the yard and faced the house. Joe turned 90° to the right. Steve turned 180° to the right. Bob turned 270° to the right. Name the object that each boy is now facing.

   Joe _____

   Steve _____

   Bob _____

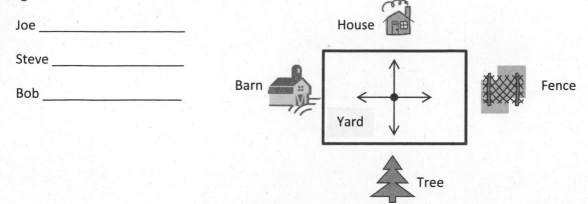

2. Monique looked at the clock at the beginning of class and at the end of class. How many degrees did the minute hand turn from the beginning of class until the end?

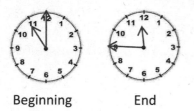

   Beginning   End

3. The skater jumped into the air and did a 360. What does that mean?

4. Mr. Martin drove away from his house without his wallet. He did a 180. Where is he heading now?

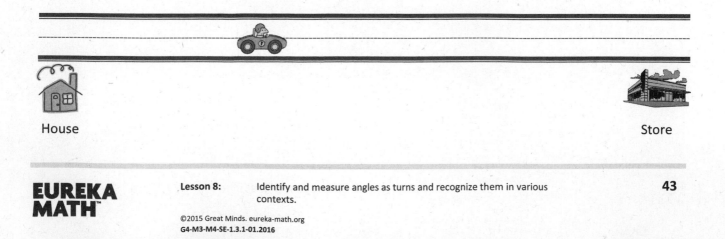

House   Store

Lesson 8:   Identify and measure angles as turns and recognize them in various contexts.

43

EUREKA MATH™

©2015 Great Minds. eureka-math.org
G4-M3-M4-SE-1.3.1-01.2016

5. John turned the knob of the shower 270° to the right. Draw a picture showing the position of the knob after he turned it.

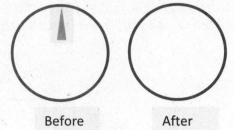

Before   After

6. Barb used her scissors to cut out a coupon from the newspaper. How many quarter-turns does she need to turn the paper in order to stay on the lines?

7. How many quarter-turns does the picture need to be rotated in order for it to be upright?

8. Meredith faced north. She turned 90° to the right, and then 180° more. In which direction is she now facing?

Lesson 8:   Identify and measure angles as turns and recognize them in various contexts.

©2015 Great Minds. eureka-math.org
G4-M3-M4-SE-1.3.1-01.2016

EUREKA
MATH™

Name _____ Date _____

1. Jill, Shyan, and Barb stood in the middle of the yard and faced the barn. Jill turned 90° to the right. Shyan turned 180° to the left. Barb turned 270° to the left. Name the object that each girl is now facing.

Jill _____

Shyan _____

Barb _____

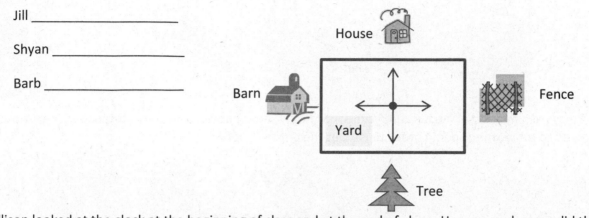

House

Barn

Fence

Yard

Tree

2. Allison looked at the clock at the beginning of class and at the end of class. How many degrees did the minute hand turn from the beginning of class until the end?

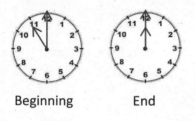

Beginning   End

3. The snowboarder went off a jump and did a 180. In which direction was the snowboarder facing when he landed? How do you know?

4. As she drove down the icy road, Mrs. Campbell slammed on her brakes. Her car did a 360. Explain what happened to Mrs. Campbell's car.

EUREKA MATH

Lesson 8:   Identify and measure angles as turns and recognize them in various contexts.

©2015 Great Minds. eureka-math.org
G4-M3-M4-SE-1.3.1-01.2016

45

5. Jonah turned the knob of the stove two quarter-turns. Draw a picture showing the position of the knob after he turned it.

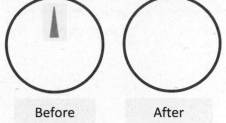

Before   After

6. Betsy used her scissors to cut out a coupon from the newspaper. How many total quarter-turns will she need to rotate the paper in order to cut out the entire coupon?

7. How many quarter-turns does the picture need to be rotated in order for it to be upright?

8. David faced north. He turned 180° to the right, and then 270° to the left. In which direction is he now facing?

Lesson 8:   Identify and measure angles as turns and recognize them in various contexts.

©2015 Great Minds. eureka-math.org
G4-M3-M4-SE-1.3.1-01.2016

EUREKA
MATH™

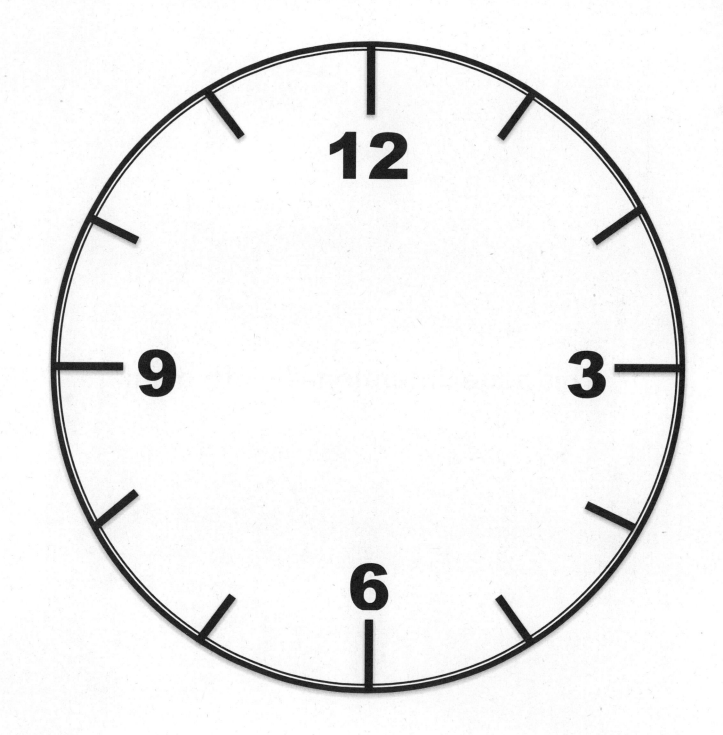

clock

Lesson 8: Identify and measure angles as turns and recognize them in various contexts.

©2015 Great Minds. eureka-math.org
G4-M3-M4-SE-1.3.1-01.2016

47

This page intentionally left blank

Name _____   Date _____

1. Complete the table.

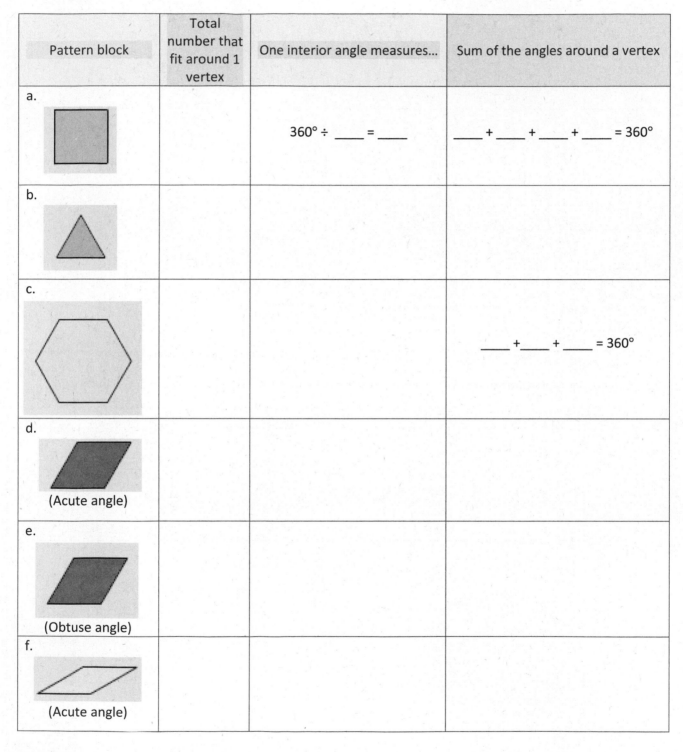

Pattern block	Total number that fit around 1 vertex	One interior angle measures...	Sum of the angles around a vertex
a.		$360° ÷$ ____ = ____	____ + ____ + ____ + ____ = 360°
b.			
c.			____ + ____ + ____ = 360°
d. (Acute angle)			
e. (Obtuse angle)			
f. (Acute angle)			

2. Find the measurements of the angles indicated by the arcs.

Pattern blocks	Angle measure	Addition sentence
a. 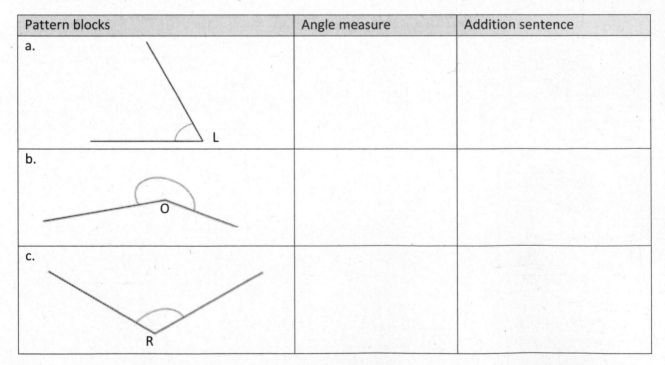		
b.		
c.		

3. Use two or more pattern blocks to figure out the measurements of the angles indicated by the arcs.

Pattern blocks	Angle measure	Addition sentence
a.		
b.		
c.		

Lesson 9:   Decompose angles using pattern blocks.

©2015 Great Minds. eureka-math.org
G4-M3-M4-SE-1.3.1-01.2016

Name _____ Date _____

Sketch two different ways to compose the given angles using two or more pattern blocks.
Write an addition sentence to show how you composed the given angle.

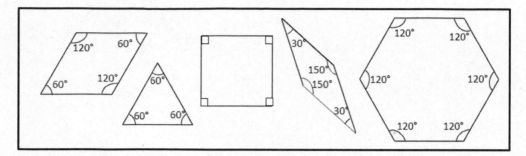

1. Points A, B, and C form a straight line.

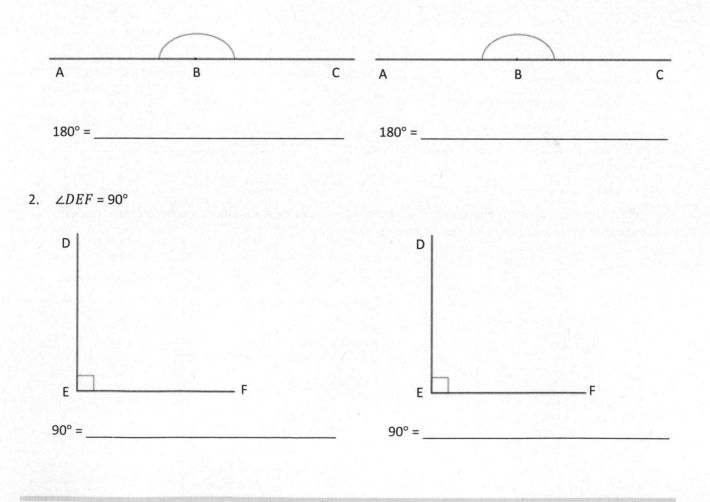

180° = _____   180° = _____

2. $\angle DEF = 90°$

90° = _____   90° = _____

3. ∠GHI = 120°

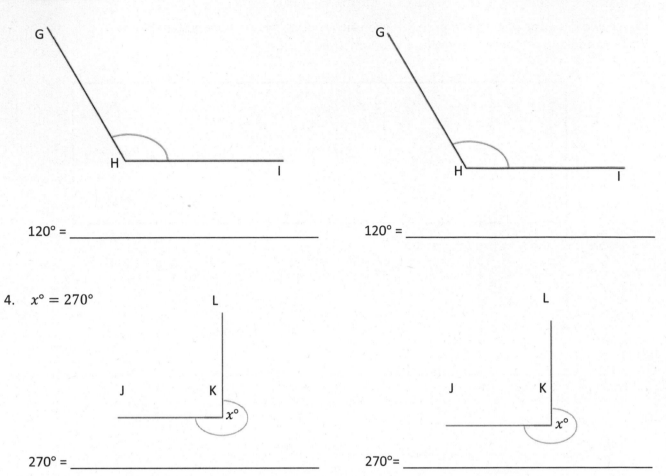

120° = _____   120° = _____

4. $x° = 270°$

120° = _____   270°= _____

270° = _____   270°= _____

5. Micah built the following shape with his pattern blocks. Write an addition sentence for each angle indicated by an arc and solve. The first one is done for you.

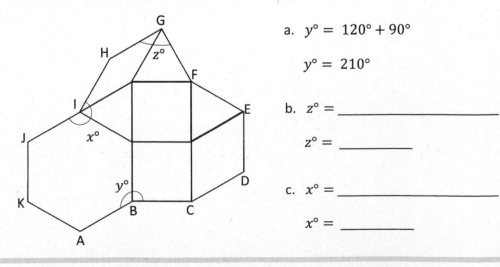

a. $y° = 120° + 90°$

$y° = 210°$

b. $z° =$ _____

$z° =$ _____

c. $x° =$ _____

$x° =$ _____

Lesson 9:   Decompose angles using pattern blocks.

©2015 Great Minds. eureka-math.org
G4-M3-M4-SE-1.3.1-01.2016

Name _____   Date _____

Write an equation, and solve for the measure of $\angle x$. Verify the measurement using a protractor.

1. $\angle CBA$ is a right angle.

45° + _____ = 90°

$x° =$ _____

2. $\angle GFE$ is a right angle.

_____ + _____ = _____

$x° =$ _____

3. $\angle IJK$ is a straight angle.

_____ + 70° = 180°

$x° =$ _____

4. $\angle MNO$ is a straight angle.

_____ + _____ = _____

$x° =$ _____

EUREKA
MATH™

Lesson 10:   Use the addition of adjacent angle measures to solve problems using a
             symbol for the unknown angle measure.

©2015 Great Minds. eureka-math.org
G4-M3-M4-SE-1.3.1-01.2016

53

Solve for the unknown angle measurements. Write an equation to solve.

5. Solve for the measurement of $\angle TRU$.
    $\angle QRS$ is a straight angle.

6. Solve for the measurement of $\angle ZYV$.
    $\angle XYZ$ is a straight angle.

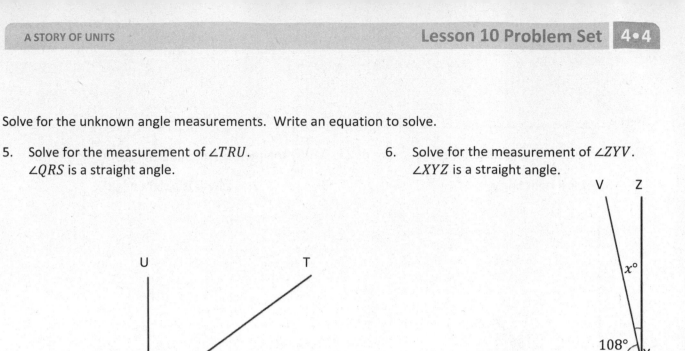

7. In the following figure, $ACDE$ is a rectangle. Without using a protractor, determine the measurement of $\angle DEB$. Write an equation that could be used to solve the problem.

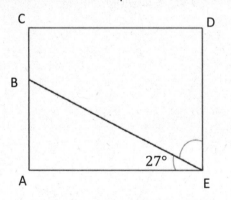

8. Complete the following directions in the space to the right.

    a. Draw 2 points: M and N. Using a straightedge, draw $\overleftrightarrow{MN}$.

    b. Plot a point O somewhere between points M and N.

    c. Plot a point P, which is not on $\overleftrightarrow{MN}$.

    d. Draw $\overline{OP}$.

    e. Find the measure of $\angle MOP$ and $\angle NOP$.

    f. Write an equation to show that the angles add to the measure of a straight angle.

Lesson 10: Use the addition of adjacent angle measures to solve problems using a symbol for the unknown angle measure.

©2015 Great Minds. eureka-math.org
G4-M3-M4-SE-1.3.1-01.2016

Name _____   Date _____

Write an equation, and solve for the measurement of ∠x. Verify the measurement using a protractor.

1. ∠DCB is a right angle.

2. ∠HGF is a right angle.

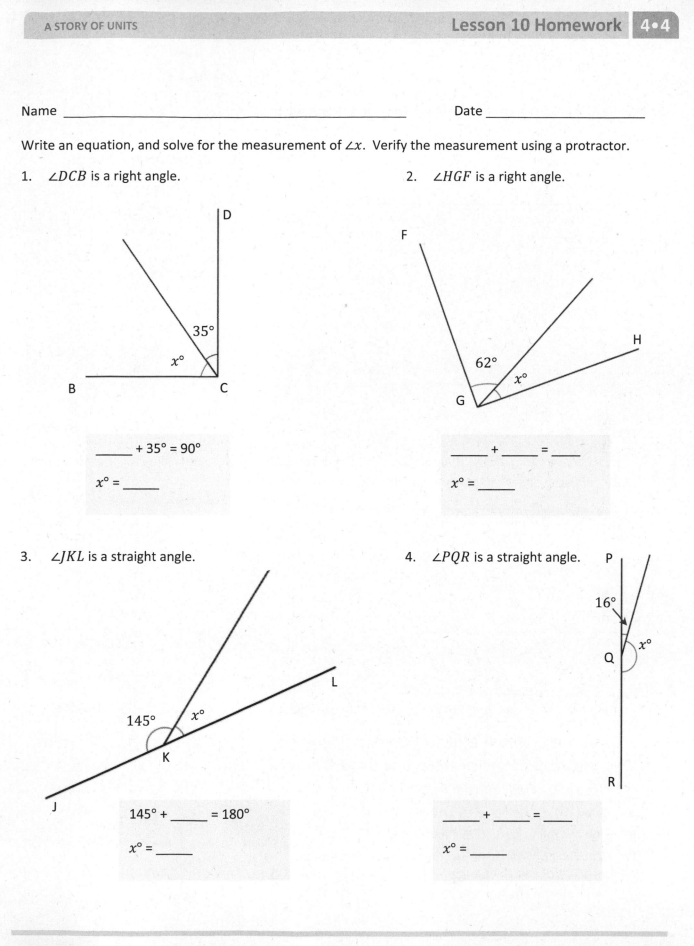

1.

_____ + 35° = 90°

$x°$ = _____

2.

_____ + _____ = _____

$x°$ = _____

3. ∠JKL is a straight angle.

4. ∠PQR is a straight angle.

3.

145° + _____ = 180°

$x°$ = _____

4.

_____ + _____ = _____

$x°$ = _____

Lesson 10:   Use the addition of adjacent angle measures to solve problems using a
             symbol for the unknown angle measure.

©2015 Great Minds. eureka-math.org
G4-M3-M4-SE-1.3.1-01.2016

Write an equation, and solve for the unknown angle measurements.

5. Solve for the measurement of ∠USW.
   ∠RST is a straight angle.

6. Solve for the measurement of ∠OML.
   ∠LMN is a straight angle.

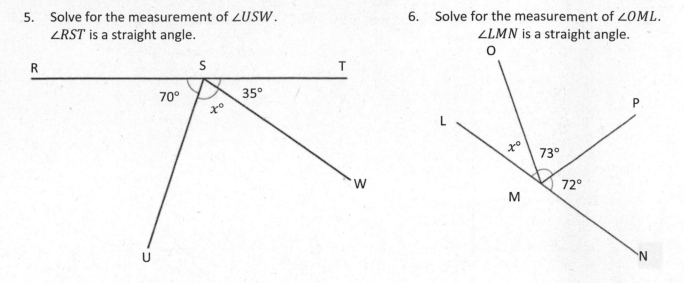

7. In the following figure, DEFH is a rectangle. Without using a protractor, determine the measurement of
   ∠GEF. Write an equation that could be used to solve the problem.

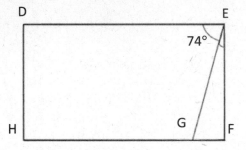

8. Complete the following directions in the space to the right.

   a. Draw 2 points: Q and R. Using a straightedge, draw $\overleftrightarrow{QR}$.

   b. Plot a point S somewhere between points Q and R.

   c. Plot a point T, which is not on $\overleftrightarrow{QR}$.

   d. Draw $\overline{TS}$.

   e. Find the measure of ∠QST and ∠RST.

   f. Write an equation to show that the angles add to the
      measure of a straight angle.

Lesson 10: Use the addition of adjacent angle measures to solve problems using a symbol for the unknown angle measure.

©2015 Great Minds. eureka-math.org
G4-M3-M4-SE-1.3.1-01.2016

EUREKA
MATH

Name _____ Date _____

Write an equation, and solve for the unknown angle measurements numerically.

1.

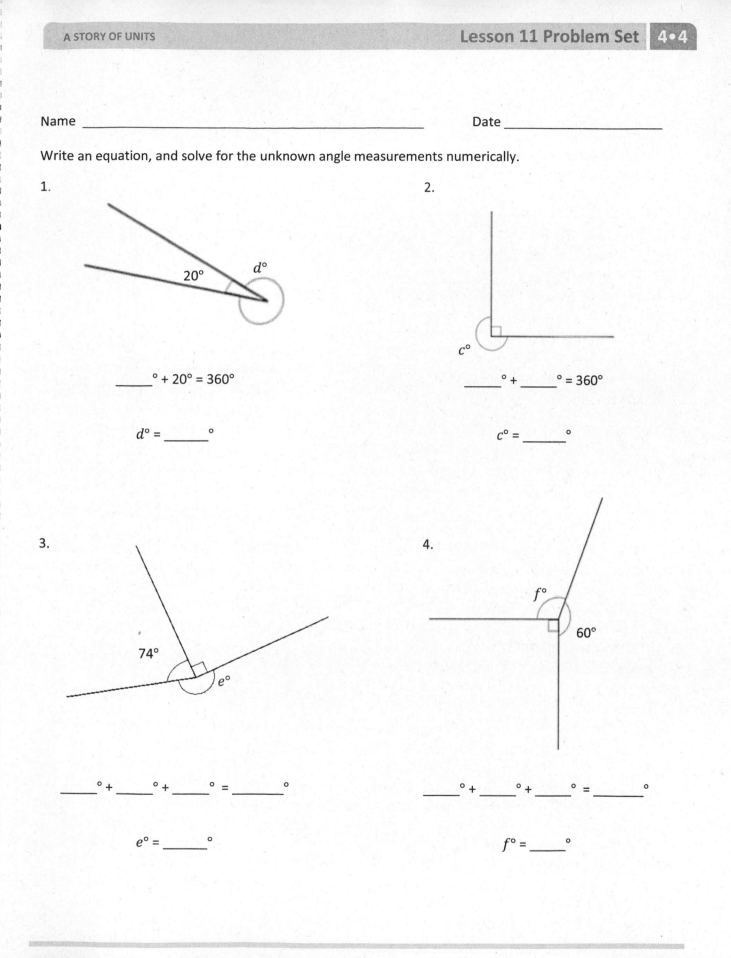

20° d°

_____° + 20° = 360°

d° = _____°

2.

c°

_____° + _____° = 360°

c° = _____°

3.

74° e°

_____° + _____° + _____° = _____°

e° = _____°

4.

f° 60°

_____° + _____° + _____° = _____°

f° = _____°

Lesson 11:   Use the addition of adjacent angle measures to solve problems using a
symbol for the unknown angle measure.

©2015 Great Minds. eureka-math.org
G4-M3-M4-SE-1.3.1-01.2016

57

Write an equation, and solve for the unknown angles numerically.

5. O is the intersection of $\overline{AB}$ and $\overline{CD}$.
    $\angle DOA$ is 160°, and $\angle AOC$ is 20°.

    $x° =$ _____ $y° =$ _____

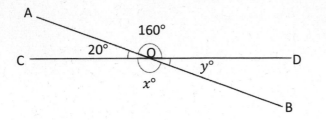

6. O is the intersection of $\overline{RS}$ and $\overline{TV}$.
    $\angle TOS$ is 125°.

    $g° =$ _____ $h° =$ _____ $i° =$ _____

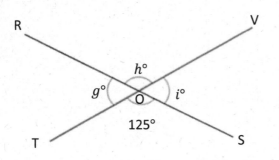

7. O is the intersection of $\overline{WX}$, $\overline{YZ}$, and $\overline{UO}$.
    $\angle XOZ$ is 36°.

    $k° =$ _____ $m° =$ _____ $n° =$ _____

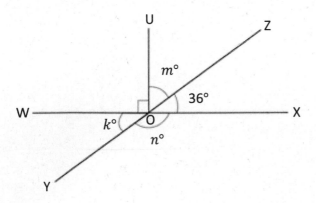

Lesson 11:   Use the addition of adjacent angle measures to solve problems using a
               symbol for the unknown angle measure.

©2015 Great Minds. eureka-math.org
G4-M3-M4-SE-1.3.1-01.2016

Name _____ Date _____

Write an equation, and solve for the unknown angle measurements numerically.

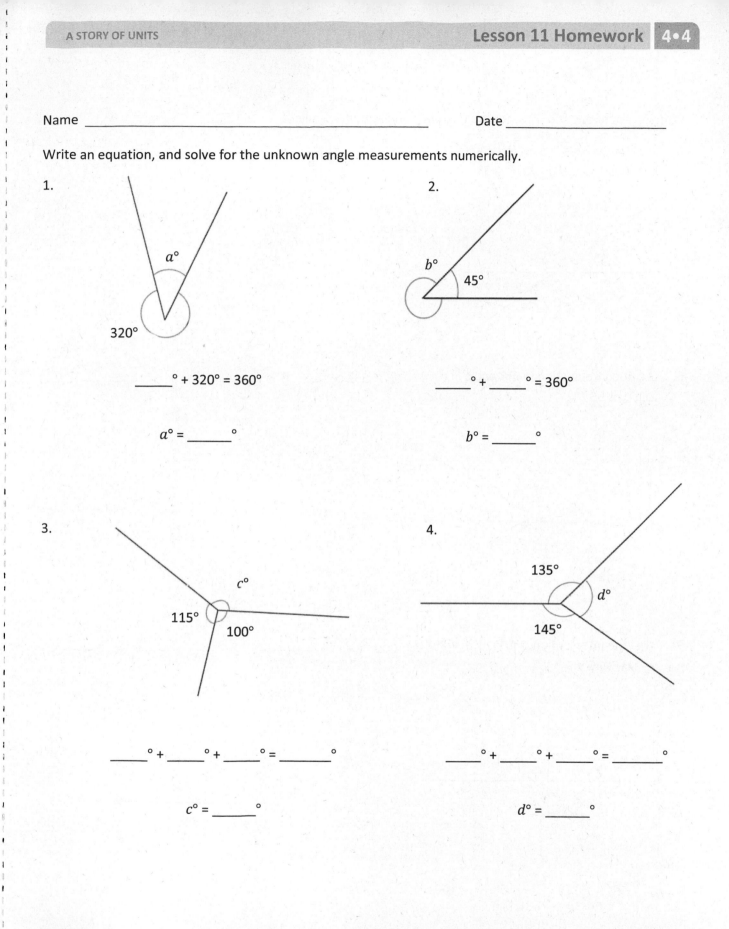

1.

a°

320°

_____° + 320° = 360°

a° = _____°

2.

b° 45°

_____° + _____° = 360°

b° = _____°

3.

c°

115° 100°

_____° + _____° + _____° = _____°

c° = _____°

4.

135° d°

145°

_____° + _____° + _____° = _____°

d° = _____°

Lesson 11: Use the addition of adjacent angle measures to solve problems using a symbol for the unknown angle measure.

59

EUREKA MATH™

©2015 Great Minds. eureka-math.org
G4-M3-M4-SE-1.3.1-01.2016

Write an equation, and solve for the unknown angles numerically.

5.   O is the intersection of $\overline{AB}$ and $\overline{CD}$.
     ∠COB is 145°, and ∠AOC is 35°.

$e° =$ _____ $f° =$ _____

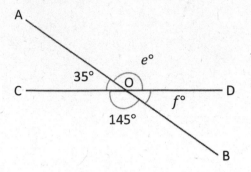

6.   O is the intersection of $\overline{QR}$ and $\overline{ST}$.
     ∠QOS is 55°.

$g° =$ _____ $h° =$ _____ $i° =$ _____

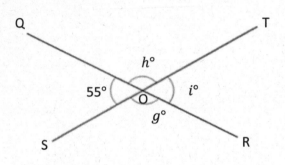

7.   O is the intersection of $\overline{UV}$, $\overline{WX}$, and $\overline{YO}$.
     ∠VOX is 46°.

$j° =$ _____ $k° =$ _____ $m° =$ _____

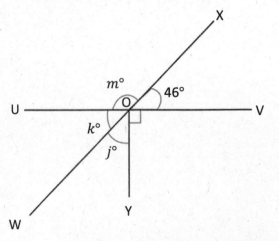

Lesson 11:   Use the addition of adjacent angle measures to solve problems using a symbol for the unknown angle measure.

©2015 Great Minds. eureka-math.org
G4-M3-M4-SE-1.3.1-01.2016

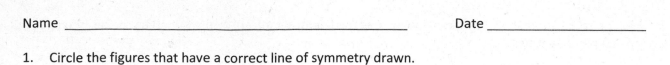

Name _____   Date _____

1. Circle the figures that have a correct line of symmetry drawn.

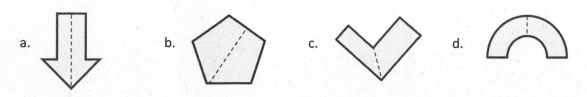

a.   b.   c.   d.

2. Find and draw all lines of symmetry for the following figures. Write the number of lines of symmetry that you found in the blank underneath the shape.

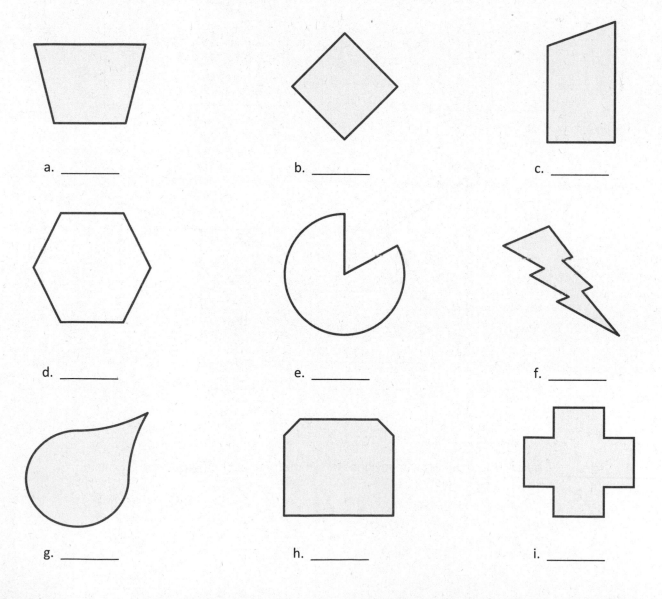

a. _____   b. _____   c. _____

d. _____   e. _____   f. _____

g. _____   h. _____   i. _____

Lesson 12:   Recognize lines of symmetry for given two-dimensional figures.   61
              Identify line-symmetric figures, and draw lines of symmetry.

©2015 Great Minds. eureka-math.org
G4-M3-M4-SE-1.3.1-01.2016

3. Half of each figure below has been drawn. Use the line of symmetry, represented by the dashed line, to complete each figure.

a.

b.

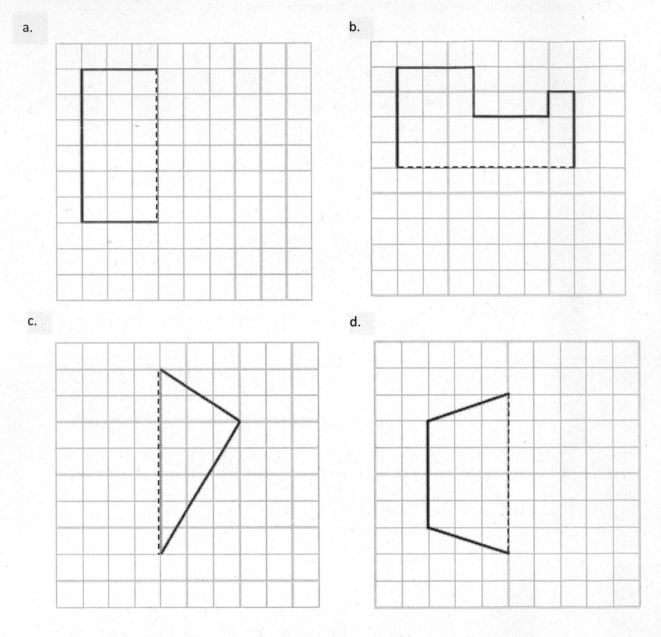

c.

d.

4. The figure below is a circle. How many lines of symmetry does the figure have? Explain.

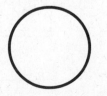

Lesson 12: Recognize lines of symmetry for given two-dimensional figures. Identify line-symmetric figures, and draw lines of symmetry.

©2015 Great Minds. eureka-math.org
G4-M3-M4-SE-1.3.1-01.2016

Name _____   Date _____

1. Circle the figures that have a correct line of symmetry drawn.

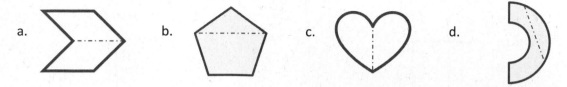

   a.   b.   c.   d.

2. Find and draw all lines of symmetry for the following figures. Write the number of lines of symmetry that you found in the blank underneath the shape.

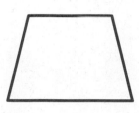   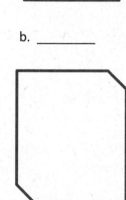   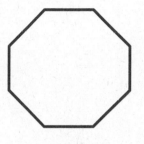

   a. _____   b. _____   c. _____

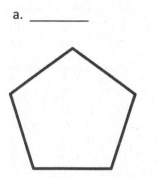      

   d. _____   e. _____   f. _____

   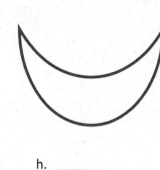

   g. _____   h. _____   i. _____

EUREKA MATH™

Lesson 12:   Recognize lines of symmetry for given two-dimensional figures.
             Identify line-symmetric figures, and draw lines of symmetry.

©2015 Great Minds. eureka-math.org
G4-M3-M4-SE-1.3.1-01.2016

63

3. Half of each figure below has been drawn. Use the line of symmetry, represented by the dashed line, to complete each figure.

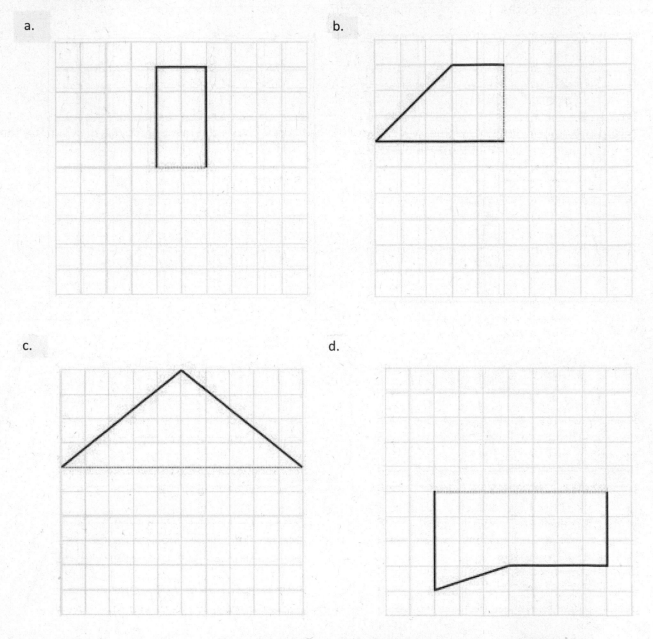

a.

b.

c.

d.

4. Is there another shape that has the same number of lines of symmetry as a circle? Explain.

Lesson 12: Recognize lines of symmetry for given two-dimensional figures. Identify line-symmetric figures, and draw lines of symmetry.

©2015 Great Minds. eureka-math.org
G4-M3-M4-SE-1.3.1-01.2016

Figure 1

Figure 2

lines of symmetry

Lesson 12:   Recognize lines of symmetry for given two-dimensional figures.
Identify line-symmetric figures, and draw lines of symmetry.

©2015 Great Minds. eureka-math.org
G4-M3-M4-SE-1.3.1-01.2016

65

This page intentionally left blank

Name _____   Date _____

Sketch of Triangle	Attributes (Include side lengths and angle measures.)	Classification	
A			
B			
C			
D			
E			
F			

Lesson 13:   Analyze and classify triangles based on side length, angle measure, or both.

©2015 Great Minds. eureka-math.org
G4-M3-M4-SE-1.3.1-01.2016

67

Name _____ Date _____

1. Classify each triangle by its side lengths and angle measurements. Circle the correct names.

	Classify Using Side Lengths	Classify Using Angle Measurements
a.	Equilateral Isosceles Scalene	Acute Right Obtuse
b.	Equilateral Isosceles Scalene	Acute Right Obtuse
c.	Equilateral Isosceles Scalene	Acute Right Obtuse
d.	Equilateral Isosceles Scalene	Acute Right Obtuse

2. △ ABC has one line of symmetry as shown. What does this tell you about the measures of ∠A and ∠C?

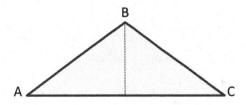

3. △ DEF has three lines of symmetry as shown.

a. How can the lines of symmetry help you to figure out which angles are equal?

b. △ DEF has a perimeter of 30 cm. Label the side lengths.

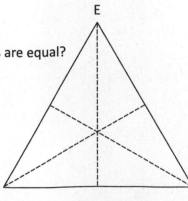

Lesson 13: Analyze and classify triangles based on side length, angle measure, or both.

©2015 Great Minds. eureka-math.org
G4-M3-M4-SE-1.3.1-01.2016

4. Use a ruler to connect points to form two other triangles. Use each point only once. None of the triangles may overlap. One or two points will be unused. Name and classify the three triangles below. The first one has been done for you.

Name the Triangles Using Vertices	Classify by Side Length	Classify by Angle Measurement
△ FJK	Scalene	Obtuse

5. a. List three points from the grid above that, when connected by segments, do not result in a triangle.

b. Why didn't the three points you listed result in a triangle when connected by segments?

6. Can a triangle have two right angles? Explain.

Lesson 13: Analyze and classify triangles based on side length, angle measure, or both.

69

©2015 Great Minds. eureka-math.org
G4-M3-M4-SE-1.3.1-01.2016

Name _____ Date _____

1. Classify each triangle by its side lengths and angle measurements. Circle the correct names.

	Classify Using Side Lengths	Classify Using Angle Measurements
a.	Equilateral Isosceles Scalene	Acute Right Obtuse
b.	Equilateral Isosceles Scalene	Acute Right Obtuse
c.	Equilateral Isosceles Scalene	Acute Right Obtuse
d.	Equilateral Isosceles Scalene	Acute Right Obtuse

2. a. △ ABC has one line of symmetry as shown. Is the measure of ∠A greater than, less than, or equal to ∠C?

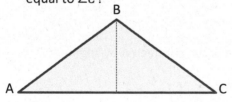

b. △ DEF is scalene. What do you observe about its angles? Explain.

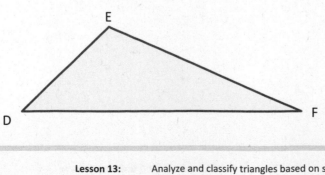

Lesson 13:   Analyze and classify triangles based on side length, angle measure, or both.

©2015 Great Minds. eureka-math.org
G4-M3-M4-SE-1.3.1-01.2016

3. Use a ruler to connect points to form two other triangles. Use each point only once. None of the triangles may overlap. Two points will be unused. Name and classify the three triangles below.

Name the Triangles Using Vertices	Classify by Side Length	Classify by Angle Measurement
△ IJK		

4. If the perimeter of an equilateral triangle is 15 cm, what is the length of each side?

5. Can a triangle have more than one obtuse angle? Explain.

6. Can a triangle have one obtuse angle and one right angle? Explain.

Lesson 13: Analyze and classify triangles based on side length, angle measure, or both.

71

©2015 Great Minds. eureka-math.org
G4-M3-M4-SE-1.3.1-01.2016

This page intentionally left blank

Name _____   Date _____

1. Draw triangles that fit the following classifications. Use a ruler and protractor. Label the side lengths and angles.

   a. Right and isosceles

   b. Obtuse and scalene

   c. Acute and scalene

   d. Acute and isosceles

2. Draw all possible lines of symmetry in the triangles above. Explain why some of the triangles do not have lines of symmetry.

Lesson 14:   Define and construct triangles from given criteria. Explore symmetry in triangles.

73

©2015 Great Minds. eureka-math.org
G4-M3-M4-SE-1.3.1-01.2016

Are the following statements true or false? Explain using pictures or words.

3.   If △ ABC is an equilateral triangle, $\overline{BC}$ must be 2 cm. True or False?

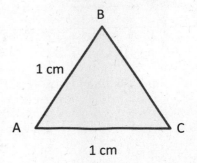

4.   A triangle cannot have one obtuse angle and one right angle. True or False?

5.   △ EFG can be described as a right triangle and an isosceles triangle. True or False?

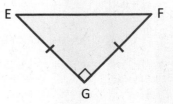

6.   An equilateral triangle is isosceles. True or False?

Extension: In △ HIJ, a = b. True or False?

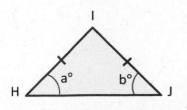

Lesson 14:   Define and construct triangles from given criteria. Explore symmetry
in triangles.

©2015 Great Minds. eureka-math.org
G4-M3-M4-SE-1.3.1-01.2016

Name _____ Date _____

1. Draw triangles that fit the following classifications. Use a ruler and protractor. Label the side lengths and angles.

   a. Right and isosceles   b. Right and scalene

   c. Obtuse and isosceles   d. Acute and scalene

2. Draw all possible lines of symmetry in the triangles above. Explain why some of the triangles do not have lines of symmetry.

Lesson 14:   Define and construct triangles from given criteria. Explore symmetry
             in triangles.

©2015 Great Minds. eureka-math.org
G4-M3-M4-SE-1.3.1-01.2016

75

Are the following statements true or false? Explain.

3.   △ ABC is an isosceles triangle. $\overline{AB}$ must be 2 cm. True or False?

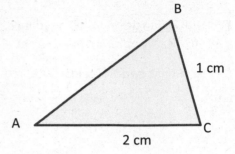

4.   A triangle cannot have both an acute angle and a right angle. True or False?

5.   △ XYZ can be described as both equilateral and acute. True or False?

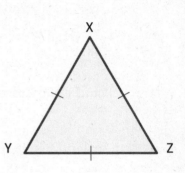

6.   A right triangle is always scalene. True or False?

Extension: In △ ABC, x = y. True or False?

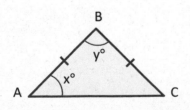

Lesson 14:   Define and construct triangles from given criteria. Explore symmetry
in triangles.

©2015 Great Minds. eureka-math.org
G4-M3-M4-SE-1.3.1-01.2016

Name _____ Date _____

Construct the figures with the given attributes. Name the shape you created. Be as specific as possible. Use extra blank paper as needed.

1. Construct quadrilaterals with at least one set of parallel sides.

2. Construct a quadrilateral with two sets of parallel sides.

3. Construct a parallelogram with four right angles.

4. Construct a rectangle with all sides the same length.

Lesson 15: Classify quadrilaterals based on parallel and perpendicular lines and the presence or absence of angles of a specified size.

©2015 Great Minds. eureka-math.org
G4-M3-M4-SE-1.3.1-01.2016

77

5. Use the word bank to name each shape, being as specific as possible.

Parallelogram	Trapezoid	Rectangle	Square

a. _____

b. _____

c. _____

d. _____

6. Explain the attribute that makes a square a special rectangle.

7. Explain the attribute that makes a rectangle a special parallelogram.

8. Explain the attribute that makes a parallelogram a special trapezoid.

Lesson 15: Classify quadrilaterals based on parallel and perpendicular lines and the presence or absence of angles of a specified size.

©2015 Great Minds. eureka-math.org
G4-M3-M4-SE-1.3.1-01.2016

EUREKA MATH™

Name _____   Date _____

1. Use the word bank to name each shape, being as specific as possible.

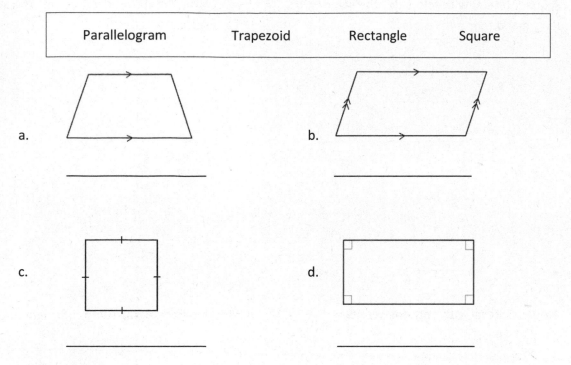

| Parallelogram | Trapezoid | Rectangle | Square |

a. _____

b. _____

c. _____

d. _____

2. Explain the attribute that makes a square a special rectangle.

3. Explain the attribute that makes a rectangle a special parallelogram.

4. Explain the attribute that makes a parallelogram a special trapezoid.

Lesson 15:   Classify quadrilaterals based on parallel and perpendicular lines and
the presence or absence of angles of a specified size.

©2015 Great Minds. eureka-math.org
G4-M3-M4-SE-1.3.1-01.2016

79

5. Construct the following figures based on the given attributes. Give a name to each figure you construct.
    Be as specific as possible.

    a.   A quadrilateral with four sides the same
         length and four right angles.

    b.   A quadrilateral with two sets of parallel
         sides.

    c.   A quadrilateral with only one set of
         parallel sides.

    d.   A parallelogram with four right angles.

Lesson 15:   Classify quadrilaterals based on parallel and perpendicular lines and
                 the presence or absence of angles of a specified size.

©2015 Great Minds. eureka-math.org
G4-M3-M4-SE-1.3.1-01.2016

Name _____   Date _____

1. On the grid paper, draw at least one quadrilateral to fit the description. Use the given segment as one segment of the quadrilateral. Name the figure you drew using one of the terms below.

| Parallelogram | Trapezoid | Rectangle |
| Square | | Rhombus |

a. A quadrilateral that has at least one pair of parallel sides.

b. A quadrilateral that has four right angles.

c. A quadrilateral that has two pairs of parallel side

d. A quadrilateral that has at least one pair of perpendicular sides and at least one pair of parallel sides.

Lesson 16:   Reason about attributes to construct quadrilaterals on square or triangular grid paper.

©2015 Great Minds. eureka-math.org
G4-M3-M4-SE-1.3.1-01.2016

81

2. On the grid paper, draw at least one quadrilateral to fit the description. Use the given segment as one segment of the quadrilateral. Name the figure you drew using one of the terms below.

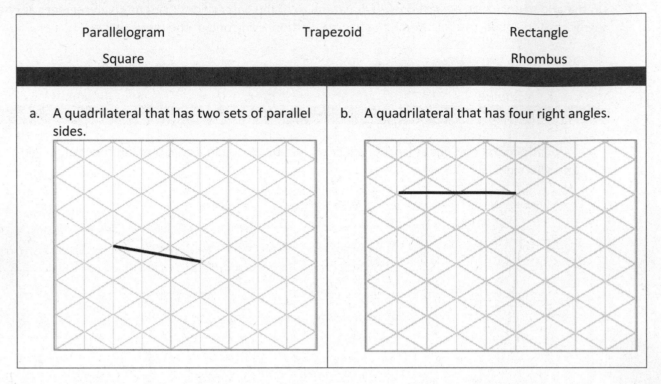

Parallelogram   Trapezoid   Rectangle

Square   Rhombus

a. A quadrilateral that has two sets of parallel sides.

b. A quadrilateral that has four right angles.

3. Explain the attributes that make a rhombus different from a rectangle.

4. Explain the attribute that makes a square different from a rhombus.

Lesson 16:   Reason about attributes to construct quadrilaterals on square or triangular grid paper.

©2015 Great Minds. eureka-math.org
G4-M3-M4-SE-1.3.1-01.2016

Name _____ Date _____

Use the grid to construct the following. Name the figure you drew using one of the terms in the word box.

1. Construct a quadrilateral with only one set of parallel sides.

    Which shape did you create?

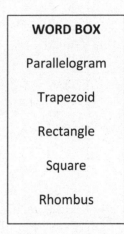

WORD BOX
Parallelogram
Trapezoid
Rectangle
Square
Rhombus

2. Construct a quadrilateral with one set of parallel sides and two right angles.

    Which shape did you create?

3. Construct a quadrilateral with two sets of parallel sides.

    Which shape did you create?

Lesson 16:   Reason about attributes to construct quadrilaterals on square or triangular grid paper.

©2015 Great Minds. eureka-math.org
G4-M3-M4-SE-1.3.1-01.2016

4. Construct a quadrilateral with all sides of equal length.
    Which shape did you create?

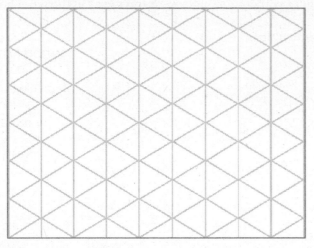

5. Construct a rectangle with all sides of equal length.
    Which shape did you create?

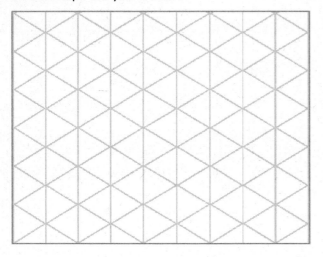

Lesson 16:   Reason about attributes to construct quadrilaterals on square or
                   triangular grid paper.

©2015 Great Minds. eureka-math.org
G4-M3-M4-SE-1.3.1-01.2016